Anonymus

Annual of the National Academy of Sciences for 1863 64

Anonymus

Annual of the National Academy of Sciences for 1863 64

ISBN/EAN: 9783742818676

Manufactured in Europe, USA, Canada, Australia, Japa

Cover: Foto ©berggeist007 / pixelio.de

Manufactured and distributed by brebook publishing software
(www.brebook.com)

Anonymus

Annual of the National Academy of Sciences for 1863 64

ANNUAL

OF THE

NATIONAL ACADEMY OF SCIENCES

FOR

1863-1864.

CAMBRIDGE:

WELCH, BIGELOW, AND COMPANY,

PRINTERS TO THE UNIVERSITY.

1865.

PREFACE.

A BY-LAW of the National Academy of Sciences (XVIII.) makes it the duty of the Secretaries to prepare an Annual, and to publish the same on the first day of each year. As but one meeting, that for organization, was held in 1863, the Secretaries were excused from the duty during that year, and the Annual now presented is intended to cover the whole period from the organization of the Academy in April, 1863, to January 1st, 1865.

WOLCOTT GIBBS.
LOUIS AGASSIZ.

CONTENTS.

I.

AN ACT

TO INCORPORATE THE NATIONAL ACADEMY OF SCIENCES.

Be it enacted by the Senate and House of Representatives of the United States of America in Congress assembled, That Louis Agassiz, Massachusetts; J. H. Alexander, Maryland; S. Alexander, New Jersey; A. D. Bache, at large; F. A. P. Barnard, at large; J. G. Barnard, United States Army, Massachusetts; W. H. C. Bartlett, United States Military Academy, Missouri; U. A. Boyden, Massachusetts; Alexis Caswell, Rhode Island; William Chauvenet, Missouri; J. H. C. Coffin, United States Naval Academy, Maine; J. A. Dahlgren, United States Navy, Pennsylvania; J. D. Dana, Connecticut; Charles H. Davis, United States Navy, Massachusetts; George Engelmann, St. Louis, Missouri; J. F. Frazer, Pennsylvania; Wolcott Gibbs, New York; J. M. Gilliss, United States Navy, Kentucky; A. A. Gould, Massachusetts; B. A. Gould, Massachusetts; Asa Gray, Massachusetts; A. Guyot, New Jersey; James Hall, New York; Joseph Henry, at large; J. E. Hilgard, at large, Illinois; Edward Hitchcock, Massachusetts; J. S. Hubbard, United States Naval Observatory, Connecticut; A. A. Humphreys, United States Army, Pennsylvania; J. L. Le Conte, United States Army, Pennsylvania; J.

Leidy, Pennsylvania; J. P. Lesley, Pennsylvania; M. F. Longstreth, Pennsylvania; D. H. Mahan, United States Military Academy, Virginia; J. S. Newberry, Ohio; H. A. Newton, Connecticut; Benjamin Peirce, Massachusetts; John Rodgers, United States Navy, Indiana; Fairman Rogers, Pennsylvania; R. E. Rogers, Pennsylvania; W. B. Rogers, Massachusetts; L. M. Rutherfurd, New York; Joseph Saxton, at large; Benjamin Silliman, Connecticut; Benjamin Silliman, Jr., Connecticut; Theodore Strong, New Jersey; John Torrey, New York; J. G. Totten, United States Army, Connecticut; Joseph Winlock, United States Nautical Almanac, Kentucky; Jeffries Wyman, Massachusetts; J. D. Whitney, California; their associates and successors duly chosen, — are hereby incorporated, constituted, and declared to be a body corporate, by the name of the National Academy of Sciences.

SECT. 2. *And be it further enacted,* That the National Academy of Sciences shall consist of not more than fifty ordinary members, and the said corporation hereby constituted shall have power to make its own organization, including its constitution, by-laws, and rules and regulations; to fill all vacancies created by death, resignation, or otherwise; to provide for the election of foreign and domestic members, the division into classes, and all other matters needful or usual in such institutions, and to report the same to Congress.

SECT. 3. *And be it further enacted,* That the National Academy of Sciences shall hold an annual meeting at such place in the United States as may be designated, and the Academy shall, whenever called upon by any department of the Government, investigate, examine, experiment, and report upon any subject of science or art, the actual expense of such investigations, examinations, experiments, and reports to be paid from appropriations which may be made for the purpose, but the Academy shall receive no compensation whatever for any services to the Government of the United States.

SOLOMON FOOTE,
President of the Senate pro tempore.

GALUSHA A. GROW,
Speaker of the House of Representatives.

Approved, March 3, 1863.

ABRAHAM LINCOLN, *President.*

II.

UNITED STATES SENATE, March 5, 1863.

SIR : —

A bill to incorporate the "National Academy of Sciences" has been introduced by me in the Senate, and, having passed through the several stages of legislation, has now become a law, under which you are one of the corporators. In the third section of this act it is enjoined, "that the National Academy of Sciences shall hold an annual meeting at such place in the United States as may be designated." In order to fulfil the injunction, and to take the first step towards the organization of the Academy, I have to request that you will be pleased to inform me, as soon as possible, at what time it will be most convenient to you to attend a meeting in New York. In naming this time, it is not necessary that you should be more specific than to give the month and part of the month.

After receiving the replies to this circular, I will select a day of meeting which will be most convenient to a majority of the members, and notify you accordingly.

I have the honor to be, with high respect,

Your most obedient servant,

HENRY WILSON.

11

III.

WASHINGTON, D. C., March 18, 1863.

SIR : —

Replies have been received to my circular letter of March 5th from more than three fifths of the members of the National Academy of Sciences named in the act of incorporation, a large majority of whom indicate no special date as more acceptable than another, leaning, however, to an early organization.

Where a choice is indicated, the dates indicated are between the last of March and beginning of July, the average being before the middle of May. May and June are excepted by some of the members.

I would therefore select, as convenient to the large part of the members, Wednesday, April 22d.

I shall, if practicable, as suggested by many of the members, be present at 11 A. M., to call the meeting to order, at the Chapel of the University of the City of New York.

I have the honor to be, with high respect,

Your most obedient servant,

HENRY WILSON.

IV.

ADDRESS

OF THE

HON. HENRY WILSON,

DELIVERED AT THE OPENING OF THE FIRST SESSION OF THE
ACADEMY, APRIL 22, 1863.

GENTLEMEN:—I hold in my hand the Act, passed in the
closing hours of the Thirty-seventh Congress, "To incorporate the National Academy of Sciences." In compliance
with many kind requests I am here to call the corporators to
order. In rising to perform this agreeable task, I crave for
a moment your indulgence.

This Act, under which you have met to organize, incor-
porates in America, and for America, a National Institution,
whose objects, ranging over the illimitable fields of science,
are limited only by the wondrous capacities of the human
intellect. Such an institution has been for years in the
thought and on the tongue of the devotees of science, but its
attainment seemed far in the future. Now it is an achieved
fact. Our country has spoken it into being, in this "dark
and troubled night" of its history, and commissioned you,
gentlemen, to mould and fashion its organization, to infuse
into it that vital and animating spirit that shall win in the
boundless domains of science the glittering prizes of achievement that will gleam forever on the brow of the nation.

When, a few months ago, a gentleman whose name is
known and honored in both hemispheres, expressed to me
the desire that an Academy of Physical Sciences should be

founded in America, and that I would at least make the effort to obtain such an act of incorporation for the scientific men of the United States, I replied, that it seemed more fitting that some statesman of ripe scholarship should take the lead in securing such a measure, but that I felt confident I could prepare, introduce, and carry through Congress a measure so eminently calculated to advance the cause of science, and to reflect honor upon our country. I promptly assumed the responsibility, and with such aid and suggestions as I could obtain, I prepared, introduced, and by personal effort with members of both Houses of Congress, carried through this act of incorporation without even a division in either House.

The suggestion was sometimes made that the nation is engaged in a fearful struggle for existence, and the moment was not well chosen to press such a measure. But I thought otherwise. I thought it just the fitting time to act. I wanted the *savans* of the old world, as they turn their eyes hitherward, to see that amid the fire and blood of the most gigantic civil war in the annals of nations, the statesmen and people of the United States, in the calm confidence of assured power, are fostering the elevating, purifying, and consolidating institutions of religion and benevolence, literature, art and science. I wanted the men of Europe, who profess to see in America the failure of republican institutions, to realize that the people of the United States, while eliminating from their system that ever-disturbing element of discord, bequeathed to them by the colonial and commercial policy of England, are cherishing the institutions that elevate man and ennoble nations. The land resounds with the tread of armies, its bright waters are crimsoned, and its fields reddened with fraternal blood. Patriotism surely demands that we strive to make this now discordant, torn, and bleeding nation one

2

and indivisible. The National Academy of Sciences will, I feel sure, be now and hereafter another element of power to keep in their orbits, around the great central sun of the Union, this constellation of sovereign commonwealths.

This act of incorporation may not be, is not, perfect. The task has been one of difficulty and delicacy. The number of members must be limited, while the most eminent men of science must be recognized, and sectional claims harmonized. If unintentional injustice has been done to any one, if mistakes have been made, time will, I trust, correct the injustice and the mistakes. Changes will surely come. "Death is in the world," and this original list of honored names will not remain long unbroken. If men of merit have been forgotten in this act of incorporation, the Academy should seize the first and every occasion to right the seeming wrong.

This Academy is destined, I trust, to live as long as the republic shall endure, and to bear upon its rolls the names of the *savans* of coming generations. Let it then advance high its standard. Let it be as inflexible as justice, and as uncompromising as truth. Let it speak with the authority of knowledge, that pretension may shrink abashed before it, and merit everywhere turn to it confident of recognition.

In the Providence of God, the Thirty-seventh Congress was summoned to the consideration of measures of transcendent magnitude. It enacted measures, empowering the government to raise hundreds of millions of dollars and millions of men, to protect the menaced life of the nation and preserve the vital spirit of freedom. It dealt with great questions of revenue and of finance. It obliterated an abhorrent system from the national capital, and engraved freedom upon every rood of the national territory. It consecrated the public domain to homesteads for the homeless and landless, and authorized the construction of a railway to unite the Atlantic and the Pacific seas. The enactment of this act to incorporate the Academy of Sciences, was not the least in the long list of acts the Thirty-seventh Congress gave to the country, which will leave their impress upon the nation for ages yet to come. It was my fortune to take a humble part in these great measures of legislation. It is a source of profound gratification to me, that, amid the pressure of public affairs, I have been enabled to contribute something to found this Academy for the advancement of the physical sciences in America. It will ever be among my most cherished recollections, that I have been permitted through your courtesy to unite with you in organizing this National Academy, which, we fondly hope, will gather around it, in the centuries yet to come, the illustrious sons of genius and of learning, whose researches will enrich the sciences, and reflect unfading lustre upon the republic.

V.

CONSTITUTION AND BY–LAWS

OF THE ACADEMY.

PREAMBLE.

EMPOWERED by the Act of Incorporation adopted by Congress, and approved by the President of the United States, on the 4th day of March, A. D. 1863, the National Academy of Sciences do enact the following Constitution and By-Laws : —

ARTICLE I.

Of Members.

SECTION 1. The members of the Academy shall be designated as Members, Honorary Members, and Foreign Associates.

SECT. 2. The Academy shall consist of the fifty members named in the Act of Incorporation, and of such others, citizens of the United States, as shall from time to time be elected to fill vacancies, in the manner hereinafter provided.

SECT. 3. Every member shall, upon his admission, take the oath of allegiance prescribed by the Senate of the United States for its own members, and, in addition thereto, an oath faithfully to discharge the duties of a member of the National Academy of Sciences to the best of his ability. He shall also subscribe the laws of the Academy.

SECT. 4. The members of the Academy shall be arranged in two Classes, according to their special studies, viz.: A, the Class of Mathematics and Physics, and B, the Class of Natural History. The corporate members may select the Class in which they desire to be arranged.

SECT. 5. The members of the Classes shall arrange themselves in Sections, by inscribing their names under one of the following heads : — Class A. *Mathematics and Physics.* Sections : — 1. Mathematics ; 2. Physics ; 3. Astronomy, Geography, and Geodesy; 4. Mechanics; 5. Chemistry. Class B. *Natural History.* Sections : — 1. Mineralogy and Geology ; 2. Zoölogy ; 3. Botany; 4. Anatomy and Physiology ; 5. Ethnology.

But the Academy retains the power of transferring a member from one Section to another.

SECT. 6. A member may be elected an honorary member of any Section by a vote of a majority of such Section.

SECT. 7. The Academy may elect fifty Foreign Associates, who shall have the privilege of attending the meetings of the Academy, and of reading and communicating papers to it, but shall take no part in its business, and shall not be subject to its assessments.

They shall be entitled to a copy of the publications of the Academy.

ARTICLE II.

Of the Officers.

SECT. 1. The officers of the Academy shall be a President, a Vice-President, a Foreign Secretary, a Home Secretary, and a Treasurer, all of whom shall be elected for

2*

a term of six years, by a majority of votes present at the first stated session after the expiration of the current terms, provided that existing officers retain their places until their successors are elected. In case of a vacancy, the election for six years shall be held in the same manner, at the next stated session after the vacancy occurs.

SECT. 2. The officers of the Classes shall be a Chairman and a Secretary, who shall be elected at each January session. The nominations shall be open, and a majority of votes shall be necessary to elect.

SECT. 3. The officers of the Academy and the Chairmen of the Classes, together with four members, two from each Class, to be annually elected by the Academy, at the January session, by a plurality of the votes, shall constitute a Council for the transaction of such business as may be assigned to them by the Constitution or the Academy.

SECT. 4. The President of the Academy, or, in case of his absence or inability to act, the Vice-President, shall preside at the meetings of the Academy and of the Council; shall name all committees, except such as are otherwise especially provided for; refer investigations required by the Government of the United States to members specially conversant with the subject, and report thereon to the Academy at its next January session; and, with the Council, shall direct the general business of the Academy.

It shall be competent for the President, in special cases, to call in the aid, upon committees, of experts, or men of remarkable attainments, not members of the Academy.

SECT. 5. The Foreign and Home Secretaries shall conduct the correspondence proper to their respective departments, advising with the President and Council in cases of doubt, and reporting their action to the Academy at its January session. It shall be the duty of the Home Secretary to give notice to the members of the place and time of all meetings, and to make known to the Council all vacancies in the list of members.

The minutes of each session shall be duly engrossed before the next stated session, under the direction of the Home Secretary.

Sect. 6. The Treasurer shall attend to all receipts and disbursements of the Academy, giving such bond, and furnishing such vouchers, as the Council may require. He shall collect all dues from members, and keep a set of books, showing a full account of receipts and disbursements. He shall present at each stated session a list of the members entitled to vote, and a general report at the January session. He shall be the custodian of the corporate seal of the Academy.

ARTICLE III.

Of the Meetings.

Sect. 1. The Academy shall hold two stated sessions in each year, — one in the city of Washington, on the 3d day of January (unless that day falls on Sunday, when the session shall be held on the succeeding Monday), and one in August, at such time and place as the Academy shall have determined upon, in private meeting, on the last day of the preceding January session.

Sect. 2. The names of the members present at each daily meeting shall be recorded in the minutes; and the members present at any meeting shall constitute a quorum for the transaction of business.

Sect. 3. Scientific meetings of the Academy, unless otherwise ordered by a majority of the members present, shall be open to the public; those for the transaction of business closed.

Sect. 4. The Academy may divide into Classes for scientific or other business. In like manner, the Classes may divide into Sections.

Sect. 5. The Classes shall meet during such periods of the stated meetings of the Academy as may be fixed by the Academy. Special meetings of a Class may be called by the Council at the request of five members of the Class.

Sect. 6. The stated meetings of the Council shall be held at the times of the stated or special meetings of the Academy. Special meetings shall be convened at the call of the President and two members of the Council, or of four members of the Council.

Sect. 7. No member who has not paid his dues shall take part in the business of the Academy.

ARTICLE IV.

Of Elections, Regulations, and Expulsions.

Sect. 1. All elections shall be by ballot, unless otherwise ordered by this Constitution; and each election shall be held separately.

Sect. 2. Whenever any election is to be held, the presiding officer shall name a Committee to conduct it, to collect the votes, count them, and report the result to the Academy. The same law shall apply in the Classes.

Sect. 3. Nominations for officers shall be made at the close of the first daily meeting of a stated session; and no candidate shall be voted for unless thus nominated.

Sect. 4. For election of members, the Council shall first decide the Class in which the vacancy shall be filled. Each Section of that Class may then select one or more candidates, after a discussion of their qualifications, and present their claims to the Class, who shall select three to be presented, in the order of their preference, to the Academy; from these three the Academy shall elect by a majority of the members present. The member elect shall be assigned to the Section in which he has been proposed. The Academy may nominate candidates in any Section which fails to propose them for itself.

SECT. 5. Every member elect shall accept his membership personally or in writing, before the close of the next stated session after the date of his election. Otherwise, on proof that the Secretary has formally notified him of his election, his name shall not be entered on the roll of members.

SECT. 6. Elections of Foreign Associates shall be conducted as follows : —

Each Section shall report to its Class, nominating a candidate whose special researches need not belong within the province of the Section, but must be comprised within the range of the Class.

From these candidates each Class shall select one name to be presented to the Academy, and from these two names the Academy, after full discussion, shall make the election, at such time as it may have previously appointed for the purpose.

SECT. 7. A diploma, with the corporate seal of the Academy and the signatures of the officers, shall be sent by the appropriate Secretary to each member on his acceptance of his membership.

SECT. 8. Resignations shall be addressed to the President and acted on by the Academy. No resignation of membership shall be accepted unless all dues have been paid.

SECT. 9. Members resigning in good standing will retain an honorary membership ; being admitted to the meetings of the Academy, but without taking part in the business. Honorary members will not be liable to assessment.

SECT. 10. If any member be absent from four consecutive stated meetings of the Academy, without communicating to the Academy a valid reason for his absence, his name shall be stricken from the roll of members.

SECT. 11. Members and officers habitually neglecting their duties shall be impeached by the Council, and at once notified thereof in writing by the Home Secretary.

SECT. 12. Impeachments of members or officers shall first be tried before the Council; which may be convened specially for such purpose. If it decides that the impeachment is proper, such impeachment shall be tried in private session before the Academy at its next stated meeting.

SECT. 13. The expulsion of a member shall be formally and publicly announced by the President at the stated session during which such expulsion shall take place.

ARTICLE V.

Of Scientific Communications, Publications, and Reports.

SECT. 1. Papers on scientific subjects may be read at the meetings of the Academy or of the Classes or Sections to which the subject belongs.

SECT. 2. Any member of the Academy may read a paper from a person who is not a member; and shall not be considered responsible for the facts or opinions expressed by the author, but shall be held responsible for the propriety of the paper.

SECT. 3. The Academy shall provide for the publication, under the direction of the Council, of Proceedings, Memoirs, and Reports.

SECT. 4. Propositions for investigations or reports shall

originate with the Classes to which the subjects belong, and be by them submitted to the Academy for approval ; except requests from the Government of the United States, which shall be acted on by the President, who will in such cases report, if necessary, at once to the Government, and to the Academy at its next stated meeting.

SECT. 5. The judgment of the Academy shall be at all times at the disposition of the Government, upon any matter of Science or Art within the limits of the subjects embraced by it.

SECT. 6. An Annual Report to be presented to Congress, shall be prepared by the President, and before its presentation submitted by him, first to the Council, and afterwards to the Academy at its January meeting.

SECT. 7. Medals and Prizes may be established, and the means of bestowing them accepted, by the Academy, upon the recommendation of the Council ; by whom all the necessary arrangements for their establishment and award shall be made.

ARTICLE VI.

Of the Property of the Academy.

SECT. 1. All investments shall be made by the Treasurer in the corporate name of the Academy, in stocks of the United States.

SECT. 2. No contract shall be binding upon the Academy, which has not been first approved by the Council.

SECT. 3. The assessments required for the support of the Academy, shall be fixed by the Academy on the recommendation of the Council.

ARTICLE VII.

Of Additions and Amendments.

Additions and Amendments to the Constitution shall be made only at a stated session of the Academy. Notice of a proposition for such a change may be given at any stated session, and shall be referred to the Council, which may amend the proposition, and shall report thereon to the Academy at its next stated session, with a recommendation that it be accepted or rejected. Its report shall be considered by the Academy in Committee of the Whole, and immediately thereafter acted on. If the addition or amendment receive two thirds of the votes present, it shall be declared adopted, and shall have the same force as the original law.

EXPLANATORY CLAUSE.

In consequence of differences of opinion in relation to the interpretation of Section 4 of Article IV. of the Constitution of the Academy, the following resolution was passed Aug. 5, 1864: —

"*Resolved,* That the Academy is of opinion that Section 4 of Art. IV. of the Constitution is to be interpreted to mean, that any Section of either Class making a nomination shall be restricted in the choice to persons eminent in the branch or branches of science understood to be included in the title of the Section."

BY-LAWS.

OF THE OFFICERS.

I. IN the absence of the Chairman or Secretary of a Class, a member shall be chosen to perform his duties temporarily, by a plurality of the *viva voce* votes, upon open nomination.

II. The accounts of the Treasurer shall be referred to an Auditing Committee of three members, to be appointed by the Academy at the meeting at which the accounts are presented ; which committee shall report before the close of that session, and shall then be discharged.

OF THE MEETINGS.

III. A Committee of Arrangements, for each stated session of the Academy, of five members, shall be appointed by the President, the Class Secretaries to be ex-officio two of the members of the Committee. This Committee shall meet not less than two weeks previous to each session. It shall be in session during the meetings, to make arrangements for the reception of the members; to arrange the business of each day; to receive the titles of papers, reports, etc. ; and to arrange the order of reading, and in general to attend to all business and scientific arrangements.

IV. At the meetings the order of business shall be as follows : —

1. Chair taken by the President, or, in his absence, the Vice-President.
2. Roll of members called by Home Secretary.
3. Report by Treasurer of members entitled to vote.
4. Minutes of the preceding meeting read and approved.
5. Stated business.
6. Reports of President, Secretaries, Treasurer, Classes, and Committees.
7. Business from Council.
8. Other business.
9. Communications from members.
10. Communications from persons not members.
11. Announcements of the death of members. Biographical notices read.
12. Rough minutes read for correction.

V. The rules of order of the Academy shall be those of the Senate of the United States, unless otherwise directed.

VI. It shall be in order for twelve members to require that any matter of business be discussed in Committee of the Whole, for amendment; the vote upon amendments to be taken in the whole Academy; and the amended proposition or propositions to be similarly voted on.

VII. The scientific meetings shall be convened at twelve o'clock, M., in order to allow time for the business meetings of the Academy, and for the meetings of Classes, Sections, and Committees.

OF ELECTIONS AND OBITUARIES.

VIII. No more than ten Foreign Associates shall be elected at any one stated session.

IX. The death of members shall be announced by the President on the last day of each stated session, when a member shall be selected by the Academy to furnish a biographical notice of the deceased at the next stated session. If such notice be not then furnished, another member shall be selected by the Academy in place of the first, and so on until the duty is performed.

X. The deaths of such eminent scientific men of the country as have taken place since the last session of the Academy shall be announced by the President. The names shall be selected by the Council.

OF SCIENTIFIC COMMUNICATIONS, PUBLICATIONS, AND REPORTS.

XI. An analysis of the memoirs and reports read in the meeting of the Classes shall be given by the Secretaries of the Classes to the Home Secretary for publication in the proceedings of the Academy. For any failure in this duty, the delinquent officer shall be impeached by the Home Secretary.

XII. The Secretaries shall receive memoirs at any time, and report the date of their reception at the next session. But no memoir shall be published unless it has been read before the Academy, Class, or Section, and ordered to be published by the Academy. Papers shall be published in the order in which they were registered, but papers which have not been sent to the Secretary within a month from the time of their reading, shall not be published without a special vote of the Academy.

XIII. Memoirs shall date in the records of the Academy from the day of their presentation to the Academy, and the order of their presentation shall be that on which they were registered, unless changed by consent of the author.

XIV. The publication of any communication to which

remonstrance is made by the Section to which the subject belongs, shall be suspended until a second time authorized by a vote of the Academy.

XV. Papers from persons not members, read before the Academy, Classes, or Sections, and intended for publication, shall be referred, at the meeting at which they are read, to a Committee of members competent to judge whether the paper is worthy of publication. Such Committees shall report to the Academy as early as practicable, and not later than the next stated session. If they do not then report, they shall be discharged, and the paper referred to another Committee.

XVI. Abstracts of papers published in the transactions of other societies or in journals may be communicated orally to the Academy; and if, on submitting any such communication to a Committee, its publication be approved, it may be ordered for publication on a vote of the Academy.

XVII. Short communications or abstracts of memoirs may be sent by any member to the Home Secretary, who shall, if requested by the author, without delay circulate them among the members.

XVIII. An Annual of the Academy shall be prepared by the Secretaries, and published on the first day of each year.

XIX. The printing of the Academy shall be under the charge of the Secretaries and the Treasurer, as a Committee of Publication, who shall report in relation thereto at each January meeting of the Academy.

XX. The Annual Report of the Academy may be accompanied by a memorial to Congress, in regard to such investigations and other subjects as may be deemed advisable, recommending appropriations therefor when necessary.

XXI. The Home Secretary shall present to the Council estimates for books and stationery, binding, &c., required for the use of the Academy.

OF THE PROPERTY OF THE ACADEMY.

XXII. The proper Secretary shall acknowledge all donations made to the Academy, and shall report them at the next stated session.

XXIII. The books, apparatus, archives, and other property of the Academy shall be deposited in some safe place in the city of Washington. A list of the articles deposited shall be kept by the Home Secretary, who is authorized to employ a clerk to take charge of them.

XXIV. A stamp corresponding to the corporate seal of the Academy shall be kept by the Secretaries, who shall be responsible for the due marking of all books and other objects to which it is applicable.

Labels or other proper marks, of similar device, shall be placed upon objects not admitting of the stamp.

OF CHANGES IN THE BY-LAWS.

XXV. Any By-Law of the Academy may be amended or repealed on the written motion of any two members, signed by them, and presented at a stated session of the Academy; provided the same shall be approved by a majority of the members present at the next stated session.

VII.

ORGANIZATION OF THE ACADEMY.

1863 – 64.

ALEXANDER DALLAS BACHE, *President.*
JAMES DWIGHT DANA, *Vice-President.*
LOUIS AGASSIZ, *Foreign Secretary.*
WOLCOTT GIBBS, *Home Secretary.*
FAIRMAN ROGERS, *Treasurer.*

COUNCIL FOR 1863.

The Officers of the Academy and the Chairmen of the Classes *ex officio.*

CHARLES HENRY DAVIS.   LEWIS M. RUTHERFURD.
JOHN TORREY.   J. PETER LESLEY.

COUNCIL FOR 1864.

The Officers of the Academy and the Chairmen of the Classes *ex officio.*

CHARLES HENRY DAVIS.   LEWIS M. RUTHERFURD.
JOHN TORREY.   J. PETER LESLEY.

OFFICERS OF THE CLASSES.

1863.

CLASS OF MATHEMATICS AND PHYSICS.

BENJAMIN PEIRCE, *Chairman.*
BENJ. A. GOULD, *Secretary.*

CLASS OF NATURAL HISTORY.

BENJAMIN SILLIMAN, SR., *Chairman.*
J. S. NEWBERRY, *Secretary.*

1864.

CLASS OF MATHEMATICS AND PHYSICS.

BENJAMIN PEIRCE, *Chairman.*
BENJ. A. GOULD, *Secretary.*

CLASS OF NATURAL HISTORY.

AUGUSTUS A. GOULD, *Chairman.*
JAMES HALL, *Secretary.*

SECTIONS.

CLASS OF MATHEMATICS AND PHYSICS.

SECTION I. *Mathematics.*

J. G. BARNARD.
H. A. NEWTON.
THEODORE STRONG.

WILLIAM CHAUVENET.
BENJAMIN PEIRCE.
JOSEPH WINLOCK.

SECTION II. *Physics.*

A. D. BACHE.	W. H. C. BARTLETT.
F. A. P. BARNARD.	A. A. HUMPHREYS.
JOSEPH HENRY.	WM. B. ROGERS.

SECTION III. *Astronomy, Geography, and Geodesy.*

STEPHEN ALEXANDER.	ALEXIS CASWELL.
J. H. C. COFFIN.	CHARLES H. DAVIS.
J. M. GILLISS.	BENJ. A. GOULD.
ARNOLD GUYOT.	LEWIS M. RUTHERFURD.

JOHN RODGERS.

SECTION IV. *Mechanics.*

J. H. ALEXANDER.	J. F. FRAZER.
J. E. HILGARD.	D. H. MAHAN.
FAIRMAN ROGERS.	JOSEPH SAXTON.

SECTION V. *Chemistry.*

WOLCOTT GIBBS.	BENJ. SILLIMAN, JR.

JOHN TORREY.

CLASS OF NATURAL HISTORY.

SECTION I. *Mineralogy and Geology.*

J. P. LESLEY.	LEO LESQUEREUX.
JAMES HALL.	J. S. NEWBERRY.
BENJ. SILLIMAN, SR.	J. D. WHITNEY.

SECTION II. *Zoölogy.*

LOUIS AGASSIZ.	JAMES D. DANA.
SPENCER F. BAIRD.	AUGUSTUS A. GOULD.

JOHN L. LECONTE.

33

SECTION III. *Botany.*

ASA GRAY.

SECTION IV. *Anatomy and Physiology.*

JEFFRIES WYMAN.   JOHN C. DALTON.

SECTION V. *Ethnology.*

VIII.

COMMITTEES OF THE ACADEMY.

I.

Committee on Weights, Measures, and Coinage.

(Appointed May 4th, 1863, at the request of the Hon. S. P. Chase, Secretary of the Treasury of the United States, dated April 24th, 1863.)

JOSEPH HENRY, *Chairman.*

J. H. ALEXANDER.　　　ARNOLD GUYOT.
FAIRMAN ROGERS.　　　BENJ. SILLIMAN, JR.
WOLCOTT GIBBS.　　　WM. CHAUVENET.
JOHN TORREY.

A. D. BACHE.　(By resolution of the Academy.)
JOHN RODGERS.　(Jan. 5th, 1864.)
L. M. RUTHERFURD.　(Jan. 5th, 1864.)

And by authority of Art. II., Sec. 4,

SAMUEL B. RUGGLES.

Mr. Henry, Chairman of the Committee on Weights, Measures, and Coinage, reported to the Academy on behalf of the Committee, January 9th, 1864, and offered the following resolution, which was adopted.

Resolved, That the Committee on Weights, Measures, and Coinage have leave to continue their labors and business now in progress, with power.

(A copy of the Report was submitted to the Secretary of the Treasury.)

II.

Committee on the Subject of Protecting the Bottoms of Iron Ships from Injury by Salt Water.

(Appointed May 9th, 1863, at the request of the Navy Department, through Rear-Admiral Davis, made May 8th, 1863.)

WOLCOTT GIBBS, *Chairman.*

BENJ. SILLIMAN, SR.   JOHN TORREY.
ROBERT E. ROGERS.   BENJ. SILLIMAN, JR.
JOHN RODGERS.   (January, 1864.)

The Committee presented a report to the Academy, January 9th, 1864, when the following resolution was passed, and the Committee was discharged.

Resolved, That the report of the Committee on the protection of the bottoms of iron ships be adopted, and that a series of experiments on the subject be undertaken by a Committee of the Academy whenever the requisite means are provided therefor.

(A copy of the Report was forwarded to the Navy Department, February 4th, 1864.)

III.

Committee to investigate and report upon the Subject of Magnetic Deviations in Iron Ships.

(Appointed May 20th, 1863, at the request of the Navy Department, made through Rear-Admiral Charles Henry Davis, May 8th, 1864.)

A. D. BACHE, *Chairman.*

JOSEPH HENRY.   BENJ. PEIRCE.
WOLCOTT GIBBS.   CHARLES H. DAVIS.
FAIRMAN ROGERS.   (May 26th, 1863.)

And by authority of Art. II., Sec. 8,

WM. P. TROWBRIDGE.

The Committee presented a Report to the Academy, January 7th, 1864, when on motion the Report with the accompanying documents was accepted and the Committee continued.

(A copy of the Report was forwarded to the Navy Department, February 11th, 1864.)

IV.

Committee to investigate and report on Saxton's Alcoholometer.

(Appointed May 25th, 1863, at the request of A. D. Bache, Superintendent of U. S. Weights and Measures, May 25th, 1863.) .

J. F. FRAZER, *Chairman.*

F. A. P. BARNARD.   WM. CHAUVENET.

JOSEPH G. TOTTEN.

The Committee reported to the Academy, January 7th, 1864, when the following resolution was passed: —

Resolved, That the words following be added to the close of the Report, viz.: "It being understood that Mr. Saxton places this invention at the disposal of the Government without any view to remuneration."

The report was then accepted, and the Committee was discharged.

V.

Committee on Wind and Current Charts and Sailing Directions.

(Appointed May 25th, 1863, at the request of the Navy Department, conveyed through Rear-Admiral C. H. Davis, May 23d, 1863, asking for an investigation and report on the subject of discontinuing the publication, in the present form, of the Wind and Current Charts and Sailing Directions.)

F. A. P. BARNARD, *Chairman.*

J. H. ALEXANDER.	ALEXIS CASWELL.
WM. CHAUVENET.	J. H. C. COFFIN.
J. F. FRAZER.	A. GUYOT.
J. E. HILGARD.	BENJ. PEIRCE.
JOSEPH WINLOCK.	

J. P. LESLEY. (June 2d, 1863.)

J. D. DANA. (July 15th, 1863.)

The Committee on Wind and Current Charts reported January 9th, 1864, when the following resolutions offered by the Committee, were adopted: —

Resolved, by the National Academy of Sciences, That in the opinion of this Academy the volumes entitled " Sailing Directions," heretofore issued to Navigators from the Naval Observatory, and the Wind and Current Charts which they are designed to illustrate and explain, embrace much which is unsound in philosophy and little that is practically useful, and that therefore these publications ought no longer to be issued in their present form.

Resolved, That the records of meteorological phenomena and of the important facts connected with terrestrial physics, which, under the direction of the Navy Department, have been accumulated at the Observatory, are capable of being turned to valuable account, and that it is eminently desirable

4

that such information should continue to be collected and subjected to careful discussion.

Resolved, That the President of the Academy be authorized and requested to communicate to the Secretary of the Navy a copy of the foregoing resolutions and of this report, as a response to the inquiry addressed to the Academy upon this subject by that officer.

The Committee was then discharged.

VI.

A Committee on National Currency.

(Appointed September 5th, 1863, at the request of the Hon. S. P. Chase, Secretary of the Treasury, made August 17th, 1863.)

JOHN TORREY, *Chairman.*

JOSEPH HENRY.

F. A. P. BARNARD.

JOSEPH SAXTON.

And by authority of Article 11, Section 4,

GEORGE C. SCHAEFFER.

January 7th, 1864. — The Committee reported successful progress, and were continued.

The following resolutions were adopted by the Academy : —

Resolved, That said Committee be empowered to communicate directly with the Secretary of the Treasury, and to take order in reference to the matters intrusted to them.

Resolved, That the President of the Academy communicate the foregoing resolution to the Hon. Secretary of the Treasury.

(Communication made January 14th, 1864.)

VII.

A Committee on the Question of Tests for the Purity of Whiskey.

(Appointed January 14th, 1864, at the request of the Acting Surgeon-General, January 5th, 1864.)

B. SILLIMAN, Jr., *Chairman.*

JOHN TORREY.

R. E. ROGERS.

J. H. ALEXANDER.

March 12th, 1864. A communication was sent from the Committee to Acting Surgeon-General I. R. Barnes, recommending that an appropriation of $ 3,500 be made to meet the expenses of the investigation.

March 25th, 1864. A letter was received from Acting Surgeon-General Barnes, stating that an appropriation of $ 3,500 had been authorized by the Secretary of War, and that Surgeon R. S. Satterlee, U. S. A., Medical Purveyor at New York, would be instructed to pay accounts for necessary purchases, etc., upon approval by the Committee.

VIII.

A Committee on the Expansion of Steam.

February 29th, 1864. The Hon. Gideon Welles, Secretary of the Navy, invited the appointment of a committee of three members of the Academy to act jointly with three members named by the Department and with three members of the Franklin Institute of Pennsylvania, for the promotion of the Mechanic Arts, to conduct, witness, and report upon experiments which may be agreed upon by the Commission on the expansion of Steam. The experiments are to be reported as early as practicable to the Department, and to be submitted also to the National Academy of Sciences for its judgment and suggestions.

March 10. The Committee of the Academy was appointed, to consist of

FAIRMAN ROGERS.   F. A. P. BARNARD.
JOSEPH SAXTON.

[The Navy Department named as its members of the joint Commission,

HORATIO ALLEN, *Chairman.*
ADMIRAL C. H. DAVIS.   B. F. ISHERWOOD.

The Franklin Institute named as its members of the joint Commission,

J. H. TOWNE.   J. V. MERRICK.
R. A. TILGHMAN.]

IX.

A Committee on Cent Coinage.

(Appointed April 11th, 1864, upon the invitation of the Hon. S. P. Chase, Secretary of the Treasury, March 30th, 1864, to examine and report upon Aluminum bronze, and other materials for the manufacture of cent coins.)

JOHN TORREY, *Chairman.*
JOSEPH HENRY.
WOLCOTT GIBBS.
F. A. P. BARNARD.
A. D. BACHE. (By request of the Department.)

The Committee reported to the Academy, August 6th, 1864. A copy of the Report was transmitted to the Hon. W. P. Fessenden, Secretary of the Treasury, August 25th, 1864.

X.

A Committee on the Explosion of the Boiler of the U. S. Steamer Chenango.

(Appointed May 2d, 1864, by the President of the Academy, under verbal authority received from the Assistant Secretary of the Navy, G. V. Fox, April 30th, 1864.)

JOHN F. FRAZER, *Chairman.*

F. ROGERS.　　　　L. M. RUTHERFURD.

The Committee reported to the Academy, August 5th, 1864, a copy of the Report having been previously forwarded to the Department.

4 *

VIII.

MEMBERS OF THE ACADEMY.

AGASSIZ, LOUIS,	Cambridge, Mass.
ALEXANDER, JOHN HENRY,	Baltimore, Md.
ALEXANDER, STEPHEN,	Princeton, N. J.
BACHE, ALEXANDER DALLAS,	Washington, D. C.
BARNARD, FREDERICK A. P.,	New York, N. Y.
BARNARD, JOHN G.,	U. S. A., Washington, D. C.
BARTLETT, WM. H. C.,	U. S. A., West Point, N. Y.
BAIRD, SPENCER F.,	Washington, D. C. Elected, August 1864.
CASWELL, ALEXIS,	Providence, R. I.
CHAUVENET, WILLIAM,	St. Louis, Mo.
COFFIN, JOHN H. C.,	U. S. N. Newport, R. I.
DALTON, JOHN CALL,	New York, N. Y. Elected, August, 1864.
DANA, JAMES DWIGHT,	New Haven, Conn.
DAVIS, CHARLES HENRY,	U. S. N., Washington, D. C.
ENGELMANN, GEORGE,	St. Louis, Mo.
FRAZER, JOHN FRIES,	Philadelphia, Penn.
GIBBS, WOLCOTT,	Cambridge, Mass.
GILLISS, JAMES MELVILLE,	U. S. N., Washington, D. C.
GOULD, BENJAMIN APTHORP,	Cambridge, Mass.
GOULD, AUGUSTUS ADDISON,	Boston, Mass.
GRAY, ASA,	Cambridge, Mass.
GUYOT, ARNOLD,	Princeton, N. J.
HALL, JAMES,	Albany, N. Y.
HENRY, JOSEPH,	Washington, D. C.

Hilgard, Julius E.,	Washington, D. C.
Humphreys, Andrew A.,	U. S. A., Washington, D. C.
Le Conte, John L.,	U. S. A., Philadelphia, Pa.
Leidy, Joseph,	Philadelphia, Penn.
Lesley, J. Peter,	Philadelphia, Penn.
Lesquereux, Leo,	Columbus, Ohio. Elected, August, 1864.
Longstreth, Miers Fisher,	Philadelphia, Penn.
Mahan, Dennis H.,	U. S. A., West Point, N. Y.
Newberry, John S.,	Cleveland, Ohio.
Newton, Hubert A.,	New Haven, Conn.
Peirce, Benjamin,	Cambridge, Mass.
Rodgers, John,	U. S. N., Washington, D. C.
Rogers, Fairman,	Philadelphia, Penn.
Rogers, Robert E.,	Philadelphia, Penn.
Rogers, William B.,	Boston, Mass.
Rutherfurd, Lewis M.,	New York, N. Y.
Saxton, Joseph,	Washington, D. C.
Silliman, Benjamin, Sr.,	New Haven, Conn.
Silliman, Benjamin, Jr.,	New Haven, Conn.
Strong, Theodore,	New Brunswick, N. J.
Torrey, John,	New York, N. Y.
Winlock, Joseph,	Cambridge, Mass.
Wyman, Jeffries,	Cambridge, Mass.
Whitney, Josiah Dwight,	San Francisco, Cal.

There are at present, January 1, 1865, two vacancies in the Academy. Of the fifty members originally appointed by Act of Congress, three have died since the first Session of the Academy, namely, Joseph S. Hubbard, Joseph G. Totten, and Edward Hitchcock. Two members by appointment, namely, Uriah A. Boyden, and John A. Dahlgren, declined to accept membership of the Academy.

IX.

FOREIGN ASSOCIATES.

The following Foreign Associates were elected at the January Meeting in 1864 : —

Sir Wm. Rowan Hamilton.
Karl Ernst Von Baer.
Michael Faraday.
J. B. Élie de Beaumont.
Sir David Brewster.
G. A. A. Plana.*
Robert Bunsen.
Friedrich Wilhelm August Argelander.
Michel Chasles.
Henry Milne-Edwards.

* Since deceased.

X.

ANNUAL REPORT OF THE PRESIDENT

FOR 1863.

ANNUAL REPORT.

NATIONAL ACADEMY OF SCIENCES,
Washington, D. C., March 28, 1864.

HON. HANNIBAL HAMLIN,
Vice-Pres. United States, President of the Senate.
HON. SCHUYLER COLFAX,
Speaker House of Representatives.

THE Constitution of the National Academy of Sciences, (Article V, Section 6,) incorporated at the Third Session of the Thirty-seventh Congress, requires that "An Annual Report, to be presented to Congress, shall be prepared by the President, and submitted by him first to the Council, and afterward to the Academy at its January meeting." In accordance with this provision, I have submitted the following report to the Council, and to the National Academy at their first stated meeting, and now present it on their behalf to Congress.

The want of an institution by which the scientific strength of the country may be brought from time to time to the aid of the government, in guiding action by the knowledge of scientific principles and experiments, has long been felt by the patriotic scientific men of the United States. No government of Europe has been willing to dispense with a body, under some name, capable of rendering such aid to the government, and in turn of illustrating the country by scientific discovery and by literary culture.

It is a remarkable fact in our annals that, just in the midst of difficulties which would have overwhelmed less resolute men, the Thirty-seventh Congress of the United States, with an elevated policy worthy of the great nation which they represented, took occasion to bring the scientific men around them in council on scientific matters, by creating the National Academy of Sciences. Such has been the way in which the public mind has been stirred before in the annals of other countries, leading to the organization of great systems of education, science, art, and literature, to be encouraged and perfected when more peaceful and prosperous times recurred. The Bill (marked A) to Incorporate the National Academy of Sciences, was passed in the Senate of the United States in February, in the House of Representatives in March, signed by the President on the 4th of March, and, as if to render it more acceptable to the men of science of the country, without opposition, from the time when unanimous consent was asked and obtained by Mr. Wilson, in the Senate, to bring in the bill, to the signature by the President.

In pursuance of the provisions of that Act, the members of the National Academy met in New York on the 22d of April, 1863, and completed their organization, renewing by their loyal oath their obligations to serve their country and its constituted authorities to the best of their abilities and knowledge, on such subjects as were embraced in their charter, and upon which they might be consulted, and adopting a Constitution and Laws which they supposed would enable them to carry on successfully the plans of Congress as sketched in the charter.

Providing for the full and deliberate consideration and arrangement of their laws by a Committee selected for their capability in such a task, the Academy adopted the laws presented to their discussion, divided into Classes and Sections for the consideration of matters of science, elected officers, (see list marked B,) and adjourned to a stated day, the 4th of January, and to Washington, the National Capital, with which they were henceforth to be connected in their membership of the National Academy of Sciences.

The first trial of the working of the Academy was to be made, and the first effort was to be through the action of a Committee on Weights and Measures, for the appointment of which, to consider the subject of the "Uniformity of Weights, Measures, and Coins, considered in relation to domestic and international commerce," the Academy had been addressed before its adjournment by the Hon. Secretary of the Treasury, S. P. Chase.

It was obvious that the only effective and prompt mode of action by members scattered over the United States, as were the fifty named in the charter, must be through committees. Action must originate with committees, and be perfected by discussion in the general meetings of the Academy, or in the classes or sections. Decisions to be finally pronounced by the entire body.

To avoid delay in reports which might be desired by the government to be promptly furnished, the President of the Academy was authorized to transmit such reports on their reception. It has not appeared to me, except, perhaps, in one case, and in that the conclusions of the Committee had not reached me, that there was occasion to present the reports until they had been discussed in the Academy itself, and the views had been adopted; especially as this was, as I have said before, a first trial of the working of our organization. One of the committees thus acting has been able to meet so often, and with so many members at a meeting, as to show that in important cases, where consultation and discussion must be had, there will be little difficulty in effecting meetings; while in most cases correspondence amply suffices for the settlement of the questions involved, and to bring out the results in the form of a report with suggestions.

It will be seen by the spirit and words of our laws, enacted by the authority of the charter, that the members of the National Academy put their time and talents at the disposal of the country in no small or stinted measure, freely, fully, by the binding authority of an oath; asking no compensation therefor but the consciousness of contributing to judicious action by the government on matters of science. The more the wealth of such men can be drawn out from the treasury of their knowledge, the richer will the nation be; and I for one do not fear that even the suggestions which may be made to Congress of subjects in which that knowledge may be most profitably employed for our country and times, will be subject to any supposed taint of self-seeking as to power or influence. Subject to the taint of supposed desire for remuneration it cannot be, by our charter, and all our laws look away from such a centre.

Since the organization of the National Academy of Sciences in April last, six committees have been appointed under the authority of Article II., Section 4, — two by application from the Treasury Department, one from the Office of United States Weights and Measures, of the same Department, and three by application from the Navy Department, or, under its authority, from the Bureau of Navigation. These applications, referring to physical, chemical, and mathematical subjects generally, have been committed to members of the Physical Class of the Academy, with a few special exceptions only. These subjects are assuredly of eminent practical value; and if the Academy, by the reports of its committees, or by their own discussions, shall give the information asked for, or shall point out the best ways of obtaining it, the members will, at the outset, have returned to the government and country the boon of their organization as a national institution. The importance of a body which can thus put the Departments and Congress on a level with the knowledge of science of the day, and by disinterested advice may keep it out of the hands of schemers, and provide the methods, intelligence, and knowledge for experimental inquiries, will thus, in the earliest days of the organization, be put to a complete test.

The subjects embraced in the references of the Departments are as follows : —

1. From the Treasury Department. Weights, Measures, and Coins, their decimalization, &c.

2. From the Navy Department. Protection of the bottoms of iron vessels from corrosion by sea-water, and from fouling.

3. From the Navy Department. Correction of the compasses of naval vessels, especially of iron vessels and of iron-clads.

4. From the Treasury Department. Saxton's new alcoholometer, intended as a substitute for the hydrometer now in use.

5. From the Navy Department. Inquiry as to the expediency of continuing, in their present form, the publication, by the Navy Department, of the Wind and Current Charts and of the Sailing Directions.

6. From the Treasury Department. Methods of protecting the national currency from being counterfeited.

The subject of weights and measures, and of coins, is undoubtedly one of the most important in the uses of common life ; and upon a right or wrong determination in regard to the system depends the convenience of the great mass of the people of a country; and upon international determinations the convenience of all engaged, directly or indirectly, in commerce and kindred pursuits.

In the report of the Secretary of the Treasury (Hon. S. P. Chase), in 1861, occurs the following sentence in regard to weights, measures and coins (page 22) : "The Secretary desires to avail himself of this opportunity to invite the attention of Congress to the importance of a uniform system and a uniform nomenclature of weights, measures, and coins, to the commerce of the world, in which the United States already so largely shares. The wisest of our statesmen have regarded the attainment of this end, so desirable in itself, as by no means impossible. The combination of the decimal system with appropriate denominations in a scheme of weights, measures, and coins, for the international uses of commerce, leaving, if need be, the separate systems of nations untouched, is certainly not beyond the reach of the daring genius and patient endeavor which gave the steam-engine and the telegraph to the service of mankind."

This Committee, No. 1, was appointed as follows: Professor Joseph Henry, Chairman, Professor J. II. Alexander, Professor Fairman Rogers, Dr. Wolcott Gibbs, Professor A. Guyot, Professor B. Silliman, Jr., Professor William Chauvenet, Dr. John Torrey, Professor A. D. Bache (appointed by resolution of Academy), Commodore John Rodgers, U. S. N., and L. M. Rutherfurd.

It is not a little strange that in our country, where, notwithstanding the capital error committed in long retaining in use foreign coins which stood in no convenient relation to the established system, the decimal system proved at once so acceptable, nothing of the kind was effected for weights and measures. It is still more strange that the antiquated and cumbrous variety of tables by which articles of different classes were bought and sold should have been retained ; that, even in our preparation of a national system intended for practical use, neither the decimalization of the weights and measures, nor the simplicity of one weight of one name, should have been adopted. The influence of great names can alone probably explain this, without justifying it.

The Committee laid out an extended scheme of reports by their members on the weights and measures of the principal countries of the world, a part of which have been already received, and are, for the present, retained in the archives of the Committee.

The discussions in the body of this Committee were very strongly in favor of the adoption of the French metrical system, but more strongly, in fact unanimously, in favor of the effort to arrive at a thorough international system, — a universal system of weights, measures, and coins, available for the general acceptance of all nations.

The Committee has received, through oral communications from the Hon. S. B. Ruggles, delegate appointed by the government of the United States to the International Statistical Congress at Berlin, authentic information as to the propositions made or adopted in that body in regard to weights, measures, and coins.

A communication, marked C, was received from the Hon. Secretary of State, and the following resolution was adopted by the Academy in regard to it : —

" *Resolved*, That the letter of the Secretary of State be referred to the Committee on Weights and Measures, with power to take such order as may in their judgment be necessary."

This Committee had several meetings during the recesses of the Academy, and finally the following report, marked

D, was submitted, and the resolution appended to it adopted by the Academy: —

" *Resolved*, That the Committee on Weights and Measures ask leave to continue their labors and business now in progress, with the power to take action."

The second Committee was appointed at the request of the Permanent Commission of the Navy Department, through the Chief of the Bureau of Navigation, on the highly important practical subject of the protection of the bottoms of iron vessels from corrosion by salt water.

The Committee consisted of Prof. W. Gibbs, Chairman, Prof. B. Silliman, Jr., Dr. John Torrey, Dr. R. E. Rogers, Prof. Benjamin Silliman, and Commodore John Rodgers, U. S. N., who, after an examination of the subject, presented to the Academy a report which was adopted on the 9th of January. They state that the methods hitherto proposed for such protection depend upon one or other of the following principles:

1st. Those which are designed to prevent or arrest, wholly or in part, the corrosion of the metal.

2d. Those intended to avoid the accumulation of living plants and animals upon the bottoms of iron ships, known technically as fouling.

The remedies for these two very distinct classes of injury to iron vessels naturally fall under the following heads: —

a. Those in which a metallic coating or alloy is employed, or those in which paints, with or without metallic oxides, are relied on.

b. The use of some poisonous substance as an ingredient of a paint or varnish, for the specific purpose of destroying the life of those plants and animals, the accumulation of which constitutes fouling.

These are discussed in the report which is hereto appended, marked E. The Committee points out that no reliable systematic experiments have been made upon the relative power of American irons to resist corrosion by sea-water, which they consider of cardinal importance. They point out also the importance of experiments on the use of oak timber as a backing to the armor of iron vessels, and are of opinion that no method yet proposed can be considered as sufficiently tested to merit a recommendation to the Department; that the question is still an open one, and that the naval and commercial interests of the country would in all probability be materially advanced by a careful and thorough experimental investigation of the whole subject.

The Secretary of the Smithsonian Institution has offered to place the laboratory under his charge at the disposal of the Committee for the purpose of investigation.

The Committee is of the opinion that no proper investigation can be made of these important subjects, unless an appropriation to defray the necessary expenses be made by the Department, or, if necessary, by Congress.

The conclusions of the Committee were adopted by the Academy in the following resolution : —

" *Resolved*, That the report of the Committee on the Coating of Iron Ships be adopted, and that a series of experiments on this subject be undertaken by a committee of the Academy, whenever the requisite means are provided therefor."

The subject, referred by the Chief of the Bureau of Navigation, by instructions from the Navy Department, of investigation of the magnetic deviation in iron ships, and of the correction of the compasses, including the correction of those of naval vessels, was referred to Committee No. 3, whose preliminary report is presented herewith, lettered F. This Committee consists of Prof. A. D. Bache, Prof. Jos.

Henry, Prof. B. Peirce, Prof. W. Gibbs, Admiral C. H. Davis, and Prof. Fairman Rogers, and Prof. W. P. Trowbridge, appointed under Art. II., Sect. 4, of the Constitution of the Academy. It was first named by the Navy Department, and the Chairman was named by the Committee, Admiral Davis having been added when the duty was transferred to the Academy.

Two important practical results have already flowed from the operations of this Committee: one, on the suggestion of the Bureau of Navigation, the taking out one of the two binnacles which were generally used in the pilot-house of the naval vessels, interfering each with the other in its use; and the correction, between April and December, of the compasses of twenty-two iron or iron-clad vessels, or of wooden vessels in which the local attraction was found to be inconvenient, from the presence of engines and boilers, of iron rigging, and other iron works.

The inconvenience and even danger resulting from the derangement of the compasses on board of many of our iron vessels have been loudly complained of to the Navy Department. The Committee adopted Airy's method for these vessels generally, and appointed Mr. A. D. Frye, of New York, who, in former years, had corrected successfully the compasses of the iron Revenue Cutters for the Treasury Department, to make the corrections. The difficulties resulting from the rapid movements to sea and port of these vessels, have sometimes rendered the effort at correction somewhat imperfect on the first trial; but a persevering application of the method has in no case failed to effect the purpose desired. The Committee has also had under successful trial a compass invented by Mr. Ritchie, of Boston, under the especial direction of the Navy Department, and a compass by Charles A. Schott, Assistant in the Coast Survey. These are referred to in the report, lettered F, of Committee No. 3. This report contains also the results of experiments on iron vessels in the course of construction, and of iron-turreted vessels, especially of the three-turreted iron-clad, the Roanoke, and of the monitor Passaic. The five compasses of the Roanoke were compared near Newport News, by swinging the vessel and noting the deviation at different points. With this report are presented nine sub-reports, as follows : —

No. 1. List of Iron-clad Vessels in Commission or Construction, as also of Iron Vessels not armored, either purchased, constructed, or being constructed.

No. 2. Report of Professor F. Rogers on Operations on U. S. Steamer Ticonderoga.

No. 3. Report of Mr. A. D. Frye.

Nos. 4, 5, 6, 7. Report by Charles A. Schott, Assistant U. S. Coast Survey, Magnetic Survey of Roanoke and Passaic.

No. 8. Drawings and Specifications of Ritchie's Fluid Compass.*

No. 9. Drawings and Specifications of Schott's Compass Arrangement.

A Committee, No. 4, on Mr. Saxton's Alcoholometer, was appointed, as follows : Professor Frazer, Chairman, Doctor Barnard, Professor Chauvenet, and General Totten. Professor W. B. Rogers was also appointed, but declined. The report, lettered G, is herewith presented. It gives a lucid description of the instrument, which was itself presented to the examination of the members of the Academy, and concludes, after a candid examination of its advantages and defects, by recommending its use to the government in place of the Tralles hydrometer, which is now employed in the col-

* Omitted by request of Mr. Ritchie, communicated through Admiral C. H. Davis.

lection of the revenue. It is much more simple, more portable, and less liable to breakage than the Tralles instrument. It was approved by the Academy on the discussion of the report, and will therefore be presented to the Treasury Department for adoption. It is so small, that the bulb and chain, which form the measuring part of the apparatus, is contained in a box of three quarters of an inch in diameter and one inch high.

The following resolution, in regard to Mr. Saxton's hydrometer, was adopted by the Academy on the 7th of January : —

"*Resolved*, That the words following be added to the close of the report, viz.: 'It being understood that Mr. Saxton places this invention at the disposal of the government without any view to remuneration.'"

A letter from Mr. Saxton, marked H, is appended to this report.

The next subject, Committee No. 5, was brought before the Academy in the following letter of the Chief of the Bureau of Navigation of the Navy Department : —

Bureau of Navigation, Navy Department,
Washington, May 23, 1863.

Sir : — I transmit herewith a copy of a letter addressed by me to the Honorable Secretary of the Navy, on the subject of discontinuing the publication, in the present form, of the " Wind and Current Charts," and " Sailing Directions " accompanying them ; and now, with the approval of the Department, I have the honor to refer the same subject to the National Academy of Sciences, for investigation and report, requesting that, on account of the expense and the public interest, it may receive early attention.

Very respectfully, your obedient servant,
CHARLES H. DAVIS,
Chief of the Bureau.

PROFESSOR A. D. BACHE, President National Academy of Sciences.

(Copy.)

Bureau of Navigation, Navy Department,
Washington, May 21, 1863.

SIR : — I have the honor to inform the Department that the Charts and Sailing Directions, published by the late Superintendent of the Observatory, at the expense of the Government, are regarded by hydrographers and scientific men as being prolix and faulty, both in matter and arrangement, to such an extent as to render the limited amount of original information which they actually contain costly and inaccessible.

I am prepared to recommend the discontinuance of the publication of these Charts and Sailing Directions. But, in order that this question of discontinuance may be decided with deliberation, I have to request permission to refer it to the National Academy of Sciences, for investigation, and report to this Department.

I am, sir, very respectfully, your obedient servant,
CHARLES H. DAVIS,
Chief of the Bureau.

HON. GIDEON WELLES, Secretary of the Navy.

The Committee appointed on this application consisted of Professor F. A. P. Barnard, Chairman, Professor J. H. Alexander, Mr. J. P. Lesley, Professor A. Caswell, Chancellor Chauvenet, Professor J. H. C. Coffin, U.S.N., Professor J. F. Frazer, Professor A. Guyot, Mr. J. E. Hilgard, Professor B. Peirce, Professor J. D. Dana, and Professor J. Winlock, U.S.N., who came to their conclusions early in October, which were adopted by the Academy after discussion, on the 9th of January, and which are expressed in the following resolutions : —

"*Resolved by the National Academy of Sciences*, That, in the opinion of this Academy, the volumes entitled ' Sailing Directions,' heretofore issued to navigators from the Naval Observatory, and the ' Wind and Current Charts ' which they are designed to illustrate and explain, embrace much which is unsound in philosophy, and little that is practically useful, and that, therefore, these publications ought no longer to be issued in their present form.

"*Resolved*, That the records of meteorological phenomena and of other important facts connected with terrestrial physics, which, under the direction of the Navy Department, have been accumulated at the Observatory, are capable of being turned to valuable account, and that it is eminently desirable that such information should continue to be collected, and subjected to careful discussion.

"*Resolved*, That the President of the Academy be authorized and requested to communicate to the Secretary of the Navy a copy of the foregoing resolutions and of this report, as a response to the inquiry addressed to the Academy upon this subject by that officer."

The report of this Committee, marked I, is appended to this Report.

It was, on motion,

"*Resolved*, That a copy of each report made on the application of the Navy Department be forwarded by the President of the Academy to the Hon. Secretary of the Navy."

The sixth Committee, appointed by request of the Treasury Department, was upon the plans presented for preventing the counterfeiting of the national currency, and consisted of Dr. Torrey, Professor Henry, Dr. Barnard, Mr. Saxton, and Professor Schaeffer, the last named being appointed by request of the Department, and under Section 4, Article II. of the Constitution of the Academy. This Committee has labored diligently and successfully in the important matters confided to them. The facts which they have developed will, by direction of the Academy, be presented confidentially to the Secretary of the Treasury. The general resolutions adopted by the Academy are as follows : —

" *Resolved*, That the Currency Committee be empowered to communicate directly with the Secretary of the Treasury, and to take order in reference to the matters intrusted to them.

" *Resolved*, That the President of the Academy communicate the foregoing resolution to the Hon. Secretary of the Treasury."

A Committee was appointed, at the first meeting, on the form of a diploma, on a corporate seal, and a stamp for books and property, which reported progress at the January meeting, and was continued.

Another Committee was appointed, to report a rule prescribing the mode of electing Foreign Associates, which reported at the January session, and was discharged.

I append to this report the minutes of the meeting of organization of the National Academy of Sciences, at New York, marked J, and of the first regular session at Washington, in January, marked K.

The draft of the Constitution and By-Laws of the Academy prepared by the Committee appointed in April, 1863, was presented and discussed in Committee of the Whole,

6

engrossed, and finally passed, as marked L, on the 6th of January.

The following papers were read at the meetings of the January session: —

Prof. Agassiz, "On the Individuality among Animals, with reference to the Questions of Varieties and Species."

Prof. B. Peirce, "On the Elements of the Mathematical Theory of Quality."

Prof. A. D. Bache, "On the Discussion of Magnetic Observations made at Girard College Observatory, in the Years 1840–45. Parts IV., V., and VI. Horizontal Force. Investigation of the Eleven-Year Period, of the Solar Diurnal Variation and Annual Inequality, and of the Influence of the Moon."

Dr. F. A. P. Barnard, "On the Force of fired Gunpowder, and the Pressure to which Heavy Guns are actually subjected in Firing."

Dr. B. A. Gould, "Reduction of the Observations of Fixed Stars made by d'Agelet at Paris during the Years 1783–85, with a Catalogue of the corresponding Mean Places referred to the Equinox of 1800."

Prof. Agassiz, "On the Metamorphoses of Fishes."

Prof. B. Peirce, "On the Saturnian System."

Dr. T. Strong, "Notes on the Parallelogram of Forces, and on Virtual Velocities."

Prof. Agassiz, "On the Geographical Distribution of Fishes, as bearing upon their Affinities and Systematic Classification."

Prof. A. D. Bache, "On the Discussion of Magnetic Observations, &c. Parts VII., VIII., and IX. Vertical Force. Investigation of the Eleven-Year Period, of the Solar Diurnal Variation and Annual Inequality, and of the Influence of the Moon."

Dr. F. A. P. Barnard, "Description of an Anemograph, designed for the University of Mississippi."

Prof. Joseph Henry, "On Materials of Combustion for Lamps in Light-houses."

Mr. L. M. Rutherfurd, "On Photographs of the Solar Spectrum."

General J. G. Barnard, "On Tangencies of Circles and Spheres."

Prof. Stephen Alexander, "On Observations of the Planet Venus, near the Times of her Inferior Conjunction, Sept. 28, 1863, and subsequently."

Prof. Stephen Alexander, "Brief Note on the Forms of Icebergs."

These papers were referred to the Committee of Publication, to take order, and to the Council, to provide the Ways and Means for publication.

The formalities of the Constitution and By-Laws in reference to Foreign Members having been fulfilled, the following were nominated and elected Foreign Members of the National Academy of Sciences: Hamilton, Baer, Faraday, Elie de Beaumont, Brewster, Plana, Bunsen, Argelander, Chasles, Milne-Edwards.

The decease of Mr. Hubbard was announced by the President, and Dr. B. A. Gould appointed to prepare a biographical notice for the next session of the Academy.

The decease of the following scientific men, not members of the Academy, was announced: Dr. Darlington, Mr. Fitz, and Major E. B. Hunt. The following members were appointed to prepare notices of their career: Mr. Torrey, Mr. Rutherfurd, and Mr. F. A. P. Barnard.

After the reading of Dr. B. A. Gould's paper "On the Reduction of the Observations of Fixed Stars made by d'Agelet at Paris, during the Years 1783-85, with a Catalogue of the corresponding Mean Places referred to the Equinox of 1800," the following resolution was unanimously adopted: —

"*Resolved*, That the Academy, impressed with the importance of a new reduction of the observations of Piazzi, presented by Mr. Gould, recommend that such reduction be made by Government at an early period."

The Council for the year was elected, as follows: Messrs. Davis, Torrey, Rutherfurd, and Lesley.

The Academy determined to meet in New Haven, next August, on the first Wednesday, at 10 A. M., at such place as may be fixed by the Committee of Arrangement.

The Committee was appointed, as follows: Messrs. B. A. Gould and Hall, Secretaries of the Classes of Mathematics and Physics; and of Natural History, Messrs. Newton, B. Siliman, jr., and Dana.

A resolution was passed, making the President of the Academy, *ex officio*, a member of all committees.

On Tuesday evening the members of the Academy were presented, by invitation, to the Hon. Secretary of the Treasury, S. P. Chase; on Thursday evening, to the Hon. Secretary of State, Wm. H. Seward; on Friday morning, to the President of the United States; on Friday evening, they came together at the residence of the President of the Academy; on Monday, visited some of the works of fortification near Washington, with Gen. Barnard; and on Tuesday, at 2½ o'clock, adjourned to the next session.

Respectfully submitted.

A. D. BACHE,
President National Academy of Sciences.

XI.

LIST OF PAPERS PRESENTED TO THE ACADEMY,

TO JANUARY 1, 1865.

1. On the Individuality among Animals, with reference to the Questions of Varieties and Species, by Louis Agassiz.
2. On the Elements of the Mathematical Theory of Quantity, by Benjamin Peirce.
3. On the Discussion of Magnetic Observations made at Girard College Observatory, in the Years 1840 – 1845. Parts IV., V., and VI. Horizontal Force. Investigation of the Eleven-year Period of the Solar Diurnal Variation and Annual Inequality, and of the Influence of the Moon, by A. D. Bache.
4. On the Force of Fired Gunpowder, and the Pressure to which Heavy Guns are actually subjected in Firing, by F. A. P. Barnard.
5. Reduction of the Observations of the Fixed Stars made by J. J. Lepaute d'Agelet at Paris, during the years 1783 – 1785, with a Catalogue of the corresponding Mean Places referred to the Equinox of 1800, by B. A. Gould.
6. On the Metamorphoses of Fishes, by Louis Agassiz.
7. On the Saturnian System, by Benjamin Peirce.
8. Notes on the Parallelogram of Forces and on Virtual Velocities, by Theodore Strong.

6*

9. On the Geographical Distribution of Fishes, as bearing upon their Affinities and Systematic Classification, by Louis Agassiz.

10. On the Discussion of Magnetic Observations made at Girard College Observatory in the Years 1840 – 1845. Part VII., VIII., and IX. Vertical Force. Investigation of the Eleven-year Period of the Solar Diurnal Variation and Annual Inequality, and of the Influence of the Moon.

11. Description of an Anemograph designed for the University of Mississippi, by F. A. P. Barnard.

12. On Materials of Combustion for Lamps in Light-houses, by Joseph Henry.

13. On Photographs of the Solar Spectrum, by Lewis M. Rutherfurd.

14. On Tangencies of Circles and Spheres, by J. G. Barnard.

15. On Observations of the Planet Venus near the Times of her Inferior Conjunction, Sept. 28, 1863, and subsequently, by Stephen Alexander.

16. Brief Note on the Forms of Icebergs, by Stephen
• Alexander.

17. Memoir of the late Henry Fitz, by Lewis M. Rutherfurd.

18. On the Distribution of certain important Diseases in the United States, by Augustus A. Gould.

19. On the Integration of differential Equations of the first Order and higher Degrees, by Theodore Strong.

20. Criticism on the Forms of Ships, by Capt. J. Cole. (Presented by Theodore Strong.)

21. On the Light visible on the Moon's Surface, and that seen adjacent to her Edge, when the Sun is either partially or totally eclipsed, by Stephen Alexander.

22. On the Influence of the Hour of the Day on the Results of Barometric Measurements of Altitudes (not read), by Arnold Guyot.
23. On Shooting Stars, by H. A. Newton.
24. A Method of determining the Errors of a Vertical Divided Circle, by Simon Newcomb. (Presented by Benj. Peirce.)
25. Considerations relative to various Phenomena presented by certain Comets, by Stephen Alexander.

XII.

EULOGY ON JOSEPH S. HUBBARD.

By B. A. GOULD.

EULOGY

ON

JOSEPH S. HUBBARD.

By B. A. GOULD.

[Read before the National Academy at New Haven, 1864, Aug. 5.]

Mr. President and Gentlemen: —

The Constitution of our Academy, like the organic law of most Academies of Science beyond the seas, provides for the tribute of a formal Biographical Notice, pronounced in open session, in commemoration of each of our number who may be removed by death. For it is no unreasonable assumption that public benefit and individual incentives may be derived from the history of any man whose scientific services have rendered him worthy of admittance to your number.

It has been the will of God that the first place in our ranks made vacant by death should be that of Joseph Stillman Hubbard, and in obedience to your instructions I am here to tell the simple story of his life; — not without a doubt of my own ability for the task, yet glad that the lot has fallen to my share, for none outside the narrow limits of his kindred could have held him dearer.

Upon our roll, Gentlemen of the Academy, are the names of venerable men, whose usefulness has extended through a period surpassing the total duration of most human lives, and side by side with these are the names of others, who were not yet cradled when the former were full of honors,

and crowned with gray hairs. The years of our eldest and
youngest member differ by more than half a century. Yet
the first summons came, not to any of the great masters in
science who give its lustre to the new gem with which an
afflicted but regenerate land would fain crown her aching
brows ; not to those who might well claim to have finished
the work on earth, which their talents and opportunities
seemed to mark out for them ; — it came to one of the
youngest in our ranks, — the forty-sixth of the original fifty
in order of age, — to one whose work seemed chiefly in the
future, and from whom we expected bright laurels for the
Academy and for America.

When in April, 1863, we assembled for the great work of
founding a National Academy, none was more hopeful, none
more buoyant, none more impressed with the magnitude and
import of our new duties, than he. It was the realization of
the dream of his maturer years, the new Atlantis of his scientific aspiration, and his heart was full of bright anticipa-
tions, tinged with all the hues which a noble enthusiasm
could bestow.

" A better Three Days for science were never spent," he
wrote to his brother ; and to his pastor in Washington, " The
inauguration of this Academy marks the most important
epoch ever witnessed by Science in America ; — we say in
the world."

In less than four months after that meeting in New York,
his generous, fervid heart had ceased to beat. He died
1863, August 16, twenty-one days before the completion of
his fortieth year.

The custom has always seemed to me an eminently proper
one, which prefaces the history of a life by some mention
and notice of ancestry. For, — whether we adopt the European notion that the ancestor ennobles his descendant by

good deeds, or the perhaps more equitable Asiatic idea that
honor flows in an ascending course, ennobling those whose
nurturing care has thus borne fruit, — the bond of lineage
may not lightly be disregarded ; and each day's experience
teaches us anew, that "men do not gather grapes of thorns
nor figs of thistles."

I may therefore say that our departed colleague drew his
origin from the early founders of our race, from that sturdy
stock which gave character to the Colony of Massachusetts
Bay, and shaped the civilization of New England.

His first American ancestor, Mr. WILLIAM HUBBARD,
came out at the age of forty in the "*Defence*" from London, in the year 1635, and soon established himself in Ipswich, Essex County, Massachusetts ; which town he represented for eight successive years, from 1638 to 1646, in the
Legislature of the Colony. In 1662, he removed to Boston,
where he died in the year 1670, aged seventy-five years,
leaving three sons, all born in England.

The eldest of these sons and second in the line of descent
was the Rev. WILLIAM HUBBARD, a man of much note in
his day. Born in 1622, he was but thirteen years old when
his father brought him to the new world. He graduated at
Harvard College in 1642, and was in 1658 ordained colleague of Rev. THOMAS COBBETT in Ipswich, where he remained as pastor until his death in 1704; his kinsman, Rev.
JOHN ROGERS, son of the President of Harvard College,
acting as his colleague during the later years of his life.

This learned and good man was one of the first historians
of the early troubles with the Indians. Two works on this
subject were published by him in 1677, and subsequently
republished in London in one volume under the title, "*The
Present State of New England.*" His "History of New
England," left by him in manuscript, is preserved in the ar-

chives of the Massachusetts Historical Society, and forms
volumes V. and VI. of their printed "Collections." In
1688, after the departure of President INCREASE MATHER
for England, he was commissioned by Governor ANDROS to
officiate as President or Rector at the Harvard Commence-
ment, being the oldest clerical Alumnus in New England;
and as there were no graduates in that year, it is recorded
in Sewall's Diary that he delivered an oration on the occasion, although this has not been transmitted to us.* His
first wife, and the mother of his children, was Margaret,
daughter of Rev. NATHANIEL ROGERS, and said to have
been the great-granddaughter of that JOHN ROGERS who
was burnt at the stake in Smithfield, 1555, — although, according to that accurate investigator, Mr. SAVAGE, this claim
is not well substantiated.

---

* That Rev. WILLIAM HUBBARD was a man of no small indepen-
dence and decision of character, may easily be inferred from his works;
but other indications of his mental and moral force arc not wanting.
In the ecclesiastical troubles of 1667, connected with the establishment
of the "Old South Church" in Boston, he took strong ground and bore
an active part; and on the passage of a vote of censure upon himself
and his colleagues in 1670, by a committee of the Legislature, he was
one of the number who answered with a protest of such ability and con-
vincing force, that the Legislature replied by an ample apology.

John Dunton, who visited him in 1686, gives [Felt, Hist. Ipswich,
p. 230] the following description of Mr. Hubbard : "The benefit of
nature and the fatigue of study have equally contributed to his emi-
nence. Neither are we less obliged to both than himself; he freely
communicates of his learning to all who have the happiness to share in
his converse. In a word, he is learned without ostentation and vanity,
and gives all his productions such a delicate turn and grace, ... that
the features and lineaments of the child make a clear discovery and
distinction of the father; yet he is a man of singular modesty, of strict
morals, and has done as much for the conversion of the Indians as
most men in New England."

The several successive generations of our colleague's ancestors seem to have been, without exception, men of moral worth, and of influence in the community.

Rev. JOHN HUBBARD, in the fourth generation, was settled in 1698 at Jamaica, Long Island, where he was distinguished by a Christian charity and tolerance remarkable for those days. His son JOHN settled in New Haven, where he served the community in the various capacities of physician, Colonel, Representative, Judge of Probate, and Judge of Common Pleas ; and his descendants have continued to reside in the vicinity of this beautiful and classic city.*

Here our colleague was born, 1823, Sept. 7 (in the ninth generation from the American founder of his family), being

* The line of descent is as follows : —

I. WILLIAM, b. in England, 1595 ; d. Ipswich, Mass., 1670.

II. Rev. WILLIAM, b. England, 1622 (H. C. 1642) ; d. Ipswich, 1794, Sept. 14 ; married Margaret, daughter of Rev. Nathaniel Rogers.

III. JOHN, a merchant of Boston, b. Ipswich, 1648 ; d. 1710, Jan. 8 ; married Ann, daughter of Gov. John Leverett.

IV. Rev. JOHN, of Jamaica, N. Y., b. Boston, 1677, Jan. 9 (H. C. 1695) ; d. 1705, Oct. 5. [See Thompson, Hist. of Long Island, 1st ed., p. 388; also Boston News Letter, No. 79, 1705, Oct. 22.]

V. Dr. JOHN, of New Haven, b. Jamaica, 1703, Nov. 30 (A. M. Yale, 1730) ; d. 1773, Oct. 30 ; married 1724, Elizabeth Stevens.

VI. Rev. JOHN, of Meriden, Conn., b. New Haven, 1727, Jan. 24 (Y. C. 1744) ; d. 1786, Nov. 18 ; married 1750, Jan. 25, Rebecca Dickerman. [See Meriden Historical Collections.]

VII. ISAAC, of Meriden, b. Meriden, 1752, Nov. 22 ; married 1782, Dec. 5, Jane, daughter of Thomas Berry.

VIII. EZRA STILES, of New Haven, b. Meriden, 1794, May 13 ; d. 1861, Aug. 20 ; married 1820, Dec. 13, Eliza, daughter of Josiah Church.

IX. JOSEPH STILLMAN, of Washington, D.C., b. New Haven, 1823, Sept. 7 (Y. C. 1843).

the second son of EZRA STILES HUBBARD, and ELIZA
CHURCH of New Haven, — parents to whom he was more
than tenderly attached, and whose declining years were
blessed by his thoughtful devotion. Of his father, I may
quote his own words written three years ago: "My father
has done his life's work well. Unable from feeble health to
live the scholar's life to which he had been destined by his
uncle, President STILES, and honoring learning next to god-
liness, he endeavored to give his children every advantage
attainable for scholarship, devoting his life, labors, and scanty
means to this one object. Precious is his memory."

From a most interesting and touching sketch of his early
life, prepared by his admirable mother, I may be permitted
to gather some of the incidents of his boyhood illustrative of
the peculiar traits of his character, — earnestness, enthusiasm, and self-forgetfulness, modified by a wholesome love of
fun and frolic, a tender susceptibility, and an affectionate
nature. From the whole account it is manifest that in
childhood as in maturer life he made for himself a place
in the hearts of all with whom he came in contact; and I
think it may be said of him with literal truth, what is so
rarely true even of good men endowed with far less force of
character, that he had not an enemy in the world.

With him, too, the old and ever new experience came to
his parents, of the early yearning of an intellectual child
for books and knowledge, and they afterwards lamented
that this dangerous tendency was not more carefully held in
check. But although the danger of over-stimulating a recep-
tive brain can hardly be exaggerated, and though the pre-
cautions of physical education were at that time comparatively disregarded among us, — I see no reason for suspicion
of any morbid precocity. I venture to make the following
extracts from the interesting accounts kindly furnished me
by his mother : —

" In his eighth year he suffered a severe course of lung fever, and for several weeks after the crisis was past seemed vacillating between life and death. After he began to convalesce, it was almost impossible to keep his active mind quiet enough to suffer the weakened frame to recover its tone. Pictures, books, toys, everything we could devise, were put into requisition to amuse him. His father saw one day in a store a curious piece of mechanism, a puzzle which he knew would delight the child, — but it was an expensive article, and he hesitated if he ought to purchase it. But a second thought of the tired, weary boy decided the question. When he put it into Joseph's hand, as he sat bolstered up in bed, the child's eyes fairly flashed with delight. Seeing him so much amused at studying its intricacies, I left him, ... and returning after a while found him utterly exhausted. He had taken the toy to pieces to ascertain its construction, and in trying to put it together again, had so used the little strength he had gained as to leave us for many days to fear a fatal result. That was ever one of his peculiarities, — not to rest till he understood the how and why of everything he saw, or at least had learned all that could be learned about it. . . . . It was about his ninth year that he began especially to develop his peculiar taste for mathematical studies and mechanics. Though he loved play dearly, and enjoyed it with zest for a little while, he had far rather spend his hours out of school in trying experiments, endeavoring to make machines, &c. . . . One of his great efforts was to make a clock. He had been attracted by seeing his father wind up the time-piece, and had begged to examine it. A day or two after I found him in his room, surrounded by a quaint collection of bits of board, pasteboard, wire, lead, &c. To the question, 'What is the tinker about now?' he replied: 'Mother, I'm going to make a clock.' I told him we must ask his father for some tools, and perhaps he would succeed; and he did succeed, — constructing a clock in all its parts, with face, hands, &c., and which went for a time, being duly mounted on the kitchen shelf, and for making which his only tools were a pair of scissors and a jackknife.

"After that, his father procured him a small chest of tools, and from that day he had full employment for every leisure hour. The attic was appropriated for his wood-work, and the back piazza for his crucibles, castings, &c. Most of his leisure time before entering college was devoted to making a telescope, which proved to be quite a good instrument, and which he sold to a gentleman from Catskill, soon after he entered college. He made also a camera-obscura, which afforded a fund of amusement to himself and his playmates, and a press for binding books. As long as his father lived he used the blank books with which the boy supplied him at this time.

"When fitting for college, while visiting some mechanic's shop, in pursuit of material or instruction, he came in contact with EBENEZER MASON, who was then one of Yale's enthusiastic astronomers, and at once there sprung up between the young man and the boy a kindly sympathy. MASON introduced the lad to his own chosen associates in study, invited him to their rooms for work, experiments, &c.; and from that day his scientific life began in earnest. Nothing could make him so happy as permission to spend the evening he could spare from daily lessons with MASON and HAMILTON SMITH; and, when in college, to be invited to watch shooting-stars or take observations with Mr. HERRICK, was the greatest boon the world could afford him. His standing in college was above mediocrity, but not what he could easily have made it. His mind was so entirely filled with his own loved department of study, that he did not value college honors enough to give the needful attention to other branches.

"In his sixteenth year Joseph determined to take a pedestrian excursion. He set out to visit an uncle residing twenty or thirty miles north of us, and his father furnished him with all he thought needful for so short a trip. He had always kept us informed of his movements when away; and when six days had passed, and we received no intelligence from him, we began to be seriously uneasy. At length a letter came, mailed in Charlestown, Mass. He had heard MASON and SMITH talk about a mechanic in Ware, who had given them much information about casting mirrors for telescopes, and had long wished to see the man for himself. So, after tarrying one night at his uncle's, he had wended his way up to Ware, and having learned all he could from the man he sought, had proceeded on foot to Charlestown, a distance of 175 miles, in order to visit Bunker Hill."

In 1843, he graduated at Yale College. For a few months he remained at home pursuing his favorite studies, mathematics and astronomy; and in the following winter he taught for a while in a classical school. Early in 1844, he went to Philadelphia, as an assistant of WALKER, who was then beginning his astronomical labors, and whose attention had been attracted by the bright promise of the earnest and gifted youth. Here the contagious zeal of WALKER added fuel to the flame. Removed for the first time from the restraining influence of home, on which he had learned unconsciously to depend, he forgot all prudent care for himself. He observed with WALKER at the High-School Observatory all night, and computed all the day, — and I need not add that his health soon gave way. From that time he was subject to a nervous excitability before unknown to him, and to an irregular action of the heart, from which he suffered much, and which finally exhausted his strength and energies, — depriving him of that vigor of constitution with which he was originally endowed, and which might have arrested the progress of his last disease.

In the autumn of 1844, Lieut. (now Major General) FRÉMONT offered Mr. HUBBARD a position in Washington as computer of the observations for latitude and longitude made on his expeditions across the Rocky Mountains, and on the Pacific coast. These completed, and the interest of Prof. BACHE, Capt. FRÉMONT, and Col. BENTON being enlisted in his behalf by his successful and meritorious labors in Philadelphia and Washington, they obtained a promise from Mr. BANCROFT, then Secretary of the Navy, that his appointment should immediately be made out for a vacancy in the corps of Professors of Mathematics in the Navy. He was commissioned 1845, May 7, and immediately assigned to duty at the Washington Observatory, of which he continued an officer during the remainder of his life. He was elected a member of the National Institute of Washington, 1845, January 14; of the Connecticut Academy of Arts and Sciences, 1849, October 24; of the American Academy of Arts and Sciences in Boston, 1850, August 15; and of the American Philosophical Society of Philadelphia, 1852, May 7.

It would be needless, gentlemen of the Academy, did taste not forbid, for me to describe to you at any length, the embarrassments of astronomers, stationed at the Washington Observatory, while under the charge of the late Superintendent. Few of you, if any, can have failed to appreciate the painful conflict between self-respect and official proprieties, — between the emotions of the scientist, jealous of his country's reputation, and of the subordinate, whose duty in an establishment under military organization demanded tacit submission and apparent acquiescence, under a mortifying or atrocious policy. The sensitive nature of WALKER found .it impossible to endure the trial; but his pupil, HUBBARD, struggled more successfully.

Would that I might with propriety express my keen sense of the deep debt of gratitude due from American science to those able and disinterested men, some of them, happily, still of our number, who bore the mortifications of their position without flinching, that they might save the national scientific institution, which it was partially within their power to protect, from becoming a source of national disgrace. They toiled earnestly and judiciously for the sake of their hope that some small portion of their labor might bear fruit, though that fruit should not be plucked by them. They struggled against obstacles which would have deterred most men, in order that the noble instruments might render some service to science, or at least fail to be made implements of national disgrace. How well they succeeded, their record bears witness; and it will bear eternal testimony to their honor, when in its own good time history shall break the seals which the present day has necessarily affixed.

A single anecdote from many which might be told, illustrative of the state of things, may perhaps be pardonable now, although it would never have been publicly mentioned by our departed colleague, nor with his permission.

Professor HUBBARD was one morning summoned to the presence of the Superintendent, who handed him a letter just received from Germany, and desired its translation. It contained the announcement of the discovery of a new comet, together with observations of its position on two successive days, — an interval of eighteen or twenty days having of course elapsed since that time.

"I. wish an Ephemeris of this comet," said he, "to be prepared without delay, for publication in the newspaper to-morrow morning." HUBBARD respectfully suggested that three observations were requisite for computing the ele-. ments, and that even should the comet be found early in the evening, the intervals between the three dates would not be well adapted for the purpose. "Confound the elements, Mr. HUBBARD!" said the Lieutenant, using some rather strong expletives; "I want none of your Elements, I only want an Ephemeris, and I wish you would compute it at once."

What the astronomer did under the embarrassing circumstances, I do not exactly know; but I suspect that the Ephemeris, which went to the *National Intelligencer*, was computed by methods neither of OLBERS nor of BESSEL!

The first published observations of HUBBARD, so far as I am aware, were those by which, on the 4th of February, 1847, he confirmed the prediction of WALKER as to the identity of *Neptune* with one of the stars observed by LALANDE, 1795, May 10. This important discovery was made almost simultaneously by PETERSEN in Altona, and by WALKER and HUBBARD in Washington, and was of the highest importance for the accurate determination of the planet's orbit. By the employment of this ancient observation, and of the perturbations computed by PEIRCE, WALKER was enabled to deduce the orbit of *Neptune* with a precision which leaves even now very little to be desired, and which surpasses that attained by any other computer to this present day.

At the Naval Observatory HUBBARD was at once placed at the Transit-Instrument, with which he observed for four months; and was transferred to the Meridian Circle in September. Nearly nine hundred transit-observations by him may be found in the volume of Washington Observations for

1845 ; and examination has shown them to possess decided value, in spite of the very unfavorable circumstances under which they were made. During the remainder of the year he was occupied with the adjustments, and in determining the Instrumental errors, of the Meridian Circle. The thorough description of this instrument and discussion of its corrections, in the volume for 1846, is from his pen ; as also is the description of the Prime-Vertical Transit. Nearly one thousand observations with the Meridian Circle in 1846, as well as the discussion already cited, give token of his activity ; but the equal labor of endeavoring to train and instruct many others, — who were assigned to duty at the several astronomical instruments by the naval routine, although not inclined to astronomical pursuits, and indeed often affected with distaste for them, — does not appear. Nor is any mention made of the careful and laborious organization and inception of a system of zone-observations, admirably devised and arranged by one of our present colleagues in connection with Professor HUBBARD, although no public acknowledgement of their services in this respect was ever made, nor indeed claimed, by either of them. According to the plan of these zone-observations (Washington Observations, 1845, App., p. 32), the micrometers of the Mural Circle, Transit-Instrument, and Meridian Circle were provided with additional declination-threads ; additional transit-threads were inserted in the field of the Mural Circle, and the micrometer of the Transit-Instrument was rotated 90° ; thus rendering it available for the measure of differences of declination. The several zones were made to overlap by 10′ in declination, and the instruments were to be employed simultaneously upon nearly the same declination, so that a portion of the stars observed by the Mural and Transit should be identical. Thus the Transit-Instrument would give standard observations of right-ascensions for an adequate number of stars in each zone swept by the Mural Circle ; while this latter would in its turn give accurate declinations for a sufficient number of stars to determine the zones observed with the Transit. The Meridian Circle, meanwhile, was to go over the same ground independently, and thus all discordances which might arise from inevitable errors of observation would be satisfactorily disposed of.

These zone-observations were begun early in 1846, and continued till 1850, and even later ; and a large amount of material was thus collected. The zones observed during 1846 with the Meridian Circle were reduced and published (under the superintendence of Mr. FERGUSON of the Washington Observatory) in 1860 ; a portion of the Mural zones for 1846 had been reduced under the superintendence of Professor COFFIN before he left the Observatory, and a considerable amount of labor had been given by HUBBARD to the reduction of the Transit-zones for the same year. With these exceptions, nothing had been done toward the reduction, on the accession of the present Superintendent in 1861 ; although in the mean time a similar investigation had been planned by Professor ARGELANDER, completely executed by him over all the practicable region south of Bessel's limit, and with a single instrument, and the results published in 1852, under the title of " Southern Zones."

But although the great labor bestowed by COFFIN and HUBBARD on the arrangement and execution of this grand scheme proved in a great degree futile, — by reason of the neglect of the observations after they were made, by the loss of some of them, and by the reckless manner in which a large proportion of the work was done, — the value of the plan and ingenuity of the arrangement remain the same. Had the valuable and delicate instruments, and the execution of the work, remained in charge of astronomers, — rather than of gentlemen, who, however gallant and accomplished in their proper calling as lieutenants and midshipmen, could not reasonably be expected to do the work of astronomers without the requisite training, and frequently much to their distaste, — had the large sums annually voted by Congress for the support of the Observatory been in part devoted to the reduction of these observations, and to the detection of the errors lurking in the observing books, — they would have conferred high honor upon American science, and indeed formed by far the noblest achievement of practical astronomy in America. As it is, it has been found necessary to reject all the zone-observations made since 1849 ; the •remainder consist of a curious combination of observations of the most delicate character and conscientious accuracy, with others which are literally beyond criticism ; and the disregard of the original plan, and total lack of system in carrying on the work with the different instruments, has in great measure defeated the scheme, which prescribed that the same region should be swept by the Transit and the Mural. Thus the zones, when reduced, do not form a complete catalogue for the region over which they extend. Moreover, it has been found necessary to determine the zero-points, both for right-ascension and for declination, of a large proportion of the zones by observations of stars made during the last two years, at an expenditure of labor quite comparable with that of the original observations of the zones, and yet exposed to all the deleterious influences which may be exerted by the unknown proper motion of the comparison-stars during an interval of from fifteen to eighteen years. The reduction of these zones has been essentially completed, so that their publication may be looked for at no distant day ; and of this work a portion of the original excellent organization, a considerable part of the earlier zones observed with the Meridian Circle, and two thirds of all the good work done with that instrument, is due to HUBBARD.

Still, in my desire to do full honor to the generous and gifted man whose loss we mourn, I may not do injustice to the living; and at the hazard of incurring the disapproval of a colleague, happily spared to us, I must add, that for an amount of intellectual labor bestowed upon this work, greater even than HUBBARD's, and for the exquisite elegance with which the observations with the Mural Circle were elaborated and made to give character and finish to the whole work, we are indebted to Professor COFFIN, whose transfer from the Observatory to the Naval Academy was productive of more advantage to the latter institution than to the one from which, unfortunately for its welfare, he was taken away in 1853. Still his influence and example were not lost, and to Professor YARNALL we owe an ample series of admirable observations with the Mural Circle, which, in connection with those of Mr. FERGUSON at the Equatorial, saved the honor of a national institution, at the time when HUBBARD was precluded by his health from observing, and after the departure of COFFIN; and have furnished valuable observations in an unbroken line from this well-equipped establishment down to the time of its resuscitation under the original founder, Capt. GILLISS.

The most valuable of HUBBARD's observations were unquestionably those with the Prime-Vertical Transit Instrument. This is essentially the counterpart of the one originally designed by STRUVE, and which has rendered such service at Pulkowa. It was thoroughly studied and mastered by HUBBARD soon after his appointment at the Observatory, and the scientific portions of the descriptions of the instrument were from the first chiefly from his pen.

It was not, however, till the beginning of 1848, a year and a half after observations with the Prime-Vertical Instrument had been commenced, that he was officially assigned to its charge. The attainment of some definite result concerning the long mooted annual parallax of *a Lyræ*, which passes within 15′ of the zenith of Washington, was an especially cherished problem. For many years he labored towards its solution, in spite of serious and most vexatious obstacles. But the maxima and minima of the annual parallax occur at seasons very unfavorable to observations in the climate and atmosphere of Washington; and it was chiefly due to this fact, that some result was not long since attained. At the regeneration of the Observatory in 1861, he was again full of hopefulness and confidence of an early solution of this favorite problem, as well as sundry others. "Your rejoicing," he wrote, "cannot exceed mine; for it is a constant gratification to see order quenching chaos, energy overriding the old slowness, and above all our own science raising her triumphant head, and banishing the old humbug." Even at that period of his domestic bereavement and loneliness, it needed only the unwonted consciousness that Astronomy might be protected at the only national Observatory in the land, to reanimate his spirits, and give him a new stimulus to exertion. The Prime-Vertical Instrument, like the others, was soon put into complete order, and the traces of early misuse thoroughly removed; and in March, 1862, he began a new series of observations of *a Lyræ*. During the period of this series HUBBARD completed an exhaustive discussion of the influence of irregularity of pivots upon the level-reading at different altitudes; — a determination of the effect and amount of flexure by comparisons of the error of collimation deduced from reversing the telescope on a star with that resulting from reversals on the image of the threads reflected from mercury in the nadir. He had re-determined the value of the level divisions, had removed some serious discordances arising from a faulty construction of the level, and had completed tables for the more convenient reduction of the observations.

This series of observations he intended to continue for several years, but an overruling Providence willed otherwise. His last observation was on the 8th of July, 1863, not sixteen months after the first. Happily he was favored with an able and skilful collaborator in Professor WILLIAM HARKNESS, and found a worthy successor. The series is continued by Professor NEWCOMB, than whom none is more competent to carry out the plans of his lamented associate, with all the success that scientific ability or earnest devotion can insure.

Professor NEWCOMB has investigated the probable error of HUBBARD's observations of a *Lyræ*, and finds that of a single observation to be but 0."155.

In the early part of the year 1849, it was my privilege to become personally acquainted with Professor HUBBARD, and to begin a friendship which knew no cloud until the last sad severance of all earthly ties. For his affectionate solicitude in time of sickness, his sympathy and support in evil days, his cordial aid in difficulty, and his encouragement in all good works, — a debt is due to his memory which words cannot express, and which, alas! this life affords no opportunity of repaying.

Without HUBBARD's cordial approval, the plan of the *Astronomical Journal* would probably not have been carried into execution; certainly it would not, at the time when it was actually begun. He aided it in every way, — by the promise of investigations for its columns — a promise amply fulfilled; — by stimulating others both to contribute and to subscribe, — by frank criticism, by generous incitement and discriminating commendation. No one could have felt a deeper interest in it than he, and of whatever service it may have rendered, a large proportion is to be credited to him alone. The earliest letter from him in my possession, dated June 8, 1849, is almost wholly devoted to a discussion of the various plans we had previously orally debated.

In the summer of 1849, these plans were essentially matured, and after discussion with BACHE, PEIRCE, HENRY, COFFIN, WALKER, CHAUVENET, and others almost equally interested, though not themselves engaged in prosecuting the same departments of inquiry, (prominent among whom were our two honored Secretaries, and the Editors of the American Journal of Science,) it was decided to give its origin a sort of national character by causing the first public suggestion to emanate from the American Association for the Advancement of Science, which held its second session at Cambridge in August, 1849. This work HUBBARD took cordially and zealously in hand. He prepared a communication, which he laid before the Association [p. 378], representing the importance of the proposed undertaking, and the services which it might render in the development of astronomy and its kindred sciences at that critical period of our national growth. At his suggestion a committee was appointed to consider the subject, and to bring it to the notice of those interested in the advancement of astronomy. He afterwards prepared a Prospectus, and labored earnestly and with effect, for its wide circulation.

The six volumes of the Journal contain more than 210 columns of valuable contributions from his pen, — and twice during the Editor's absence from the country did HUBBARD assume the control and editorship.

8 *

The first extended computation of Professor HUBBARD consisted in the determination of the zodiacs of all the known asteroids, except the four previously published in Germany. In November, 1848, he presented to the Smithsonian Institution the Zodiacs of *Vesta, Astrea, Hebe, Flora,* and *Metis;* and to the first volume of the Astronomical Journal, he contributed those of *Hygea, Parthenope,* and *Clio,* making the list complete up to that time. That of *Egeria* followed, soon after his satisfactory determination of the elements; and although he published no others, it was his intention as well as endeavor to prepare the zodiac for each successively discovered asteroid. These zodiacs give for each planet, — as suggested by GAUSS, and computed by him for *Ceres, Pallas,* and *Juno,* — the northern and southern limits of its geocentric position for each right-ascension, and enable us in many cases to draw immediate inferences as to the possible identity of any recorded star with the planet in question. It is much to be desired that the series of asteroid-zodiacs should be completed, and a key thus furnished for the solution of many interesting questions of identity, which have occurred in the past, and must present themselves hereafter.

None of you, Gentlemen, can fail to recall the magnificent spectacle exhibited by the great Comet of 1843. Through the early evenings of March, it trailed like a gorgeous banner of flame across the Western sky, the first visitant of its kind within the memories of many a full grown man, and rekindling the awe and wonder of those, whose impressions of the cometic glories of 1807, 1811, and 1819 had become dimmed by time. Its magnificent train extended at nightfall nearly parallel with the horizon through an arc of some 40°, rivaling the later, though perhaps equally splendid, manifestation of the great Comet of 1858. So great indeed was its brilliancy while in close proximity to the sun, that it attracted the attention of the public at high noon in various parts of North and South America both on the day of its perihelion, and on the day following. It was seen at 11 o'clock on the morning of the 27th, at Conception, and measurements of its distance from the sun were made on the 28th, both in Maine and in Mexico; the tail being visible to the length of a full degree, at 3 o'clock in the afternoon of that day. The attempts of astronomers to satisfy the observations led to results singularly diverse. Only one characteristic of the orbit seemed beyond question, — the extreme smallness of the perihelion distance. The close resemblance of its parabolic elements to those deduced by HENDERSON for the Comet of 1668, could not fail to attract attention, and the elements obtained by PEIRCE from the very unsatisfactory observations of the Comet of 1689, which have come down to us, exhibit also a decided similarity. Both CAPOCCI and CLAUSSEN, believing in its identity. with both, found themselves able to satisfy the observations by an ellipse of seven years period. ENCKE, WALKER, and ANDERSON found that the observations could be closely represented by a hyperbolic orbit, — BOGUSLAWSKI in Breslau advocated a period of $147\frac{1}{4}$ years, — WALKER finally decided in favor of an ellipse of $21\frac{7}{8}$ years, — while LAUGIER and MAUVAIS in Paris, NICOLAI in Mannheim, and others, found the probabilities strongly in favor of the period of 175 years, — which I cannot but believe to be the true one.

This magnificent object fired the zeal of HUBBARD, already fascinated as he was with astronomical study and imbued with the spirit of research. He was within five months of graduation at Yale, and, from that time, he looked forward to a thorough and decisive investigation of the path of this comet, as his most favorite problem. And although some six years elapsed before he found it within his power to begin the long-desired research, he then prosecuted it with an earnestness which showed no loss of interest or of enthusiasm.

In December, 1849, he published the first part of this masterly discussion of the Orbit of the Great Comet of 1843, — an investigation begun only a few months before, but hastened for the sake of an early contribution to the Astronomical Journal. This paper occupied a part of eight numbers, the conclusion appearing in July, 1852. It seems to me safe to say that the orbit of no comet of long period has been more thoroughly and exhaustively treated than this. All observations of the comet, of whatever kind, whether before published or obtained from the manuscripts of astronomers, were subjected to rigorous scrutiny, and were winnowed with a painstaking fidelity which would have surpassed the patience of most men. Especially were the very important sextant-observations, made in the daytime on the 28th of February by Captain CLARKE, at Portland, Maine, and by Mr. BOWRING, at Chihuahua, discussed with extreme care, and made, after sundry corrections, to exert an important influence upon the resultant orbit.

First forming normal places by the aid of one of the approximate parabolas at hand, HUBBARD computed elliptic elements by the ordinary Gaussian method, and thus obtained new normals. Determining for these the coefficients of the variations of the elements relatively to the variations of the geocentric co-ordinates, — and, for the sake of control, both by BESSEL's method and by that of GOETZE, he deduced the variations required for satisfying the new normals, and thus arrived at a second set of elements.

Repeating the process, and computing ten new equations

of condition for new normal places, he obtained a third and fourth ellipse, the latter by the assignment of weights to the several normals. The amount of outstanding error was thus reduced to a very small quantity, and the orbit was sufficiently accurate to correct the sextant-observations, and decide sundry points left ambiguous by the observer. Thus he found which limit of the sun had been compared with the comet at Chihuahua, and was able to make the assumption of an error of two minutes in one of the recorded times of observation, and thus both to render the observations accordant, and to show their value. In a similar way the untrustworthiness of another sextant-observation was made manifest, and thus prevented from vitiating the computations. The errors of two sextant-measurements in each place were thus shown to lie within the limits of good observation, and the aid of these very important auxiliaries secured.

The disturbing forces were computed for each of the six large planets for each fourth day during the period of the comet's visibility, and with the series of osculating elements thus obtained, he determined the discordances of every accessible observation. Here, as everywhere in HUBBARD's work, we find the indication of his scrupulous care in controlling his computations by the independent employment of different formulas, and of the tact by which he adapted various methods to his purpose; this peculiarity, as well as his exquisite elegance in the mechanical arrangement, and the beauty of his chirography reminding one continually of ENCKE, many of whose scientific characteristics seem equally to have belonged to HUBBARD, — though the fulness of years and opportunities happily accorded to the accomplished astronomer of Berlin were denied to our departed associate. True to his nature, he computed all the anomalies and radius-vectors in duplicate, once by means of a manuscript table to supply the reductions needed for NICOLAI's formulas, which proved more convenient than BARKER's table for an orbit of so small a perihelion distance, and then again by means of the Besselian reduction of the parabola to an ellipse.

New equations of condition were now formed, sixty-six in number, — weights were empirically assigned to each, and a fifth system of elements thus found which absolutely represented the Portland observation, and satisfied the two Chihuahua altitudes so admirably that the greatest discordance of the five amounted to but 37″, while the probable error of a normal place amounted to 16″. Separating the observations made with a ring-micrometer from those obtained by the filar micrometer, he was able to assign more accurate weights to the several measurements of each coordinate, and found, as might have been anticipated, that the probable error with the ring-micrometer did not much exceed that with the filar micrometer for differences of right ascension, while it proved to be nearly in the ratio of 7 to 10 for differences of declination.

By a repetition of the process, after assigning carefully computed weights, as above mentioned, to sixty-five normal equations of condition, HUBBARD obtained by the method of least squares a sixth system of elements, which gave the best possible representation to the entire series of observations, and reduced the probable error of a normal place to less than 13″.

Here the investigation might well have rested; for the effect of terms of the second order, both in the perturbations and in the comparisons, might fairly be considered as removed, and the sums of the squares of the residuals were a minimum. But HUBBARD was not content to leave any

investigation, where there seemed an opportunity of prosecuting it further with success; and since the incorporation
of observations made with the ring-micrometer had increased the probable errors of the results, and since the'
series with the filar micrometer extended through the whole
period of visibility excepting the observations by daylight,
he passed on to still another determination from the filar-
micrometer observations alone combined with the sextant-
observations of February. 28. From these he constructed
eighty-three new equations of condition, determined a
seventh series of elements, reducing the probable error
of a single normal to less than $8''.5$, and assigned for each
element its probable error. The period corresponding to
these final elements was somewhat more than five hundred
years, and it became a problem of much interest to determine to what extent the resultant period might be varied
consistently with the probable limits of errors of observation. This HUBBARD solved most thoroughly by an ingeni-
ous method of determining the variations of each of the
elements, of the probable errors, and of each normal place,
as a function of the variation of the eccentricity. So that
by substituting in these expressions the change of eccentricity corresponding to any suspected period, a few minutes of
figuring will give us the corresponding elements, the probable error of normal places, and the individual discordances
of observations. This substitution he carried out himself
for the period of one hundred and seventy-five years, and
found that it implied a probable error of $11''\frac{1}{3}$ for a single
observation, and no individual discordances beyond the limit
of reasonable error; although, to be sure, a certain "rate"
seems indicated on this assumption by the earlier observations. The limits of periodic time consistent with its observed geocentric path were thus shown to be extremely

wide; and HUBBARD closed by suggesting that the want of coincidence between the centre of gravity and the centre of apparent condensation, as well as the operation of polar forces in the comet itself, might perhaps modify deductions drawn without consideration of these possible influences.

I have not hesitated, Gentlemen of the Academy, to describe this valuable memoir with a minuteness of detail quite unsuitable for a popular address; both because its masterly completeness and elegance render it a model investigation of its class, and because these qualities were so characteristic of our late colleague that a somewhat minute description seemed well adapted to exhibit his habits of mind and mode of research. Preserved in the Library of Yale College are three quarto volumes containing the actual numerical computations, — all executed with marvellous neatness and a beauty of penmanship approaching the elegance of copperplate engraving, — all arranged in due order, and in the form most convenient for reference, and all bearing the strong impress of the man. Indeed, in every one of his manuscripts we may see the reflection of his own cultivated and tasteful mind, in which there was no slovenly corner, or ill-finished record.

HUBBARD's next investigation of magnitude was upon Biela's comet. Four quarto volumes, filled with neat figures, lie before me as I write, containing his researches concerning the orbit of this most interesting body. They are a priceless and treasured memento of our departed friend, which I owe to the thoughtful kindness of his family; and it is not improbable that four or five years hence they may facilitate the discovery of the origin and nature of the mysterious transformations which this singular comet has undergone, and may aid in the detection of the unknown laws controlling its physical structure.

It is known to you all that Biela's comet, as it is generally called, is one of short period, performing its entire revolution in about 6¾ years. It was first seen in 1772, by MONTAIGNE, who made three or four imperfect and untrustworthy estimates of position, and it was observed four times, quite unsatisfactorily, by MESSIER. In 1806, it was detected by PONS; and the general resemblance of its orbit to the approximate one deduced for the comet of 1772, attracted immediate attention. BESSEL and GAUSS computed elliptic orbits on the supposition of identity. The latter found the apparent path as well represented by an ellipse of 4¾ years as by his best parabola, thus suggesting the probability that there had been six intermediate returns. The places observed in 1772 were, however, not so well satisfied by an ellipse of so short a major axis, and therefore while the hypothesis of identity seemed plausible, it could hardly be considered probable. It was not until 1826 that the comet was seen again. In that year it was independently discovered both on the 27th February, by VON BIELA, an Austrian captain, on duty at the fortress of Josephstadt, and by GAMBART in Marseilles, ten days later. Upon the first computation of the orbit, each recognized the identity of the comet with that of 1806, and the true length of the period became manifest.

The next return, in 1832, was successfully predicted by astronomers; at the following one in 1839, it was not discovered; and in the winter of 1845 – 6, a predicted return was for the second time observed. But here an unexpected and anomalous phenomenon was exhibited. The comet, which was detected at the close of November, was before the end of December seen to be double, and the two components became apparently farther and farther apart, until, at the end of March, their distance from one another amounted to more than 14′.

9

It was of course immediately maintained by some that an explosion had occurred, and it became a question of great interest to all astronomers, when, how, and through what agency the separation had been brought about. And yet another curious circumstance was this : — that whereas the northern and preceding component was at first so decidedly the fainter of the two as to receive the name of the " companion," while the southerly one was regarded as the comet proper ; — yet this companion, or northerly component, gradually increased in brilliancy, until about the time of perihelion-passage, surpassing the primary nucleus for several days, and then again diminishing in relative brightness so long as observations could be made.

HUBBARD, who had observed this comet at Washington early in January, 1846, had been deeply impressed with these inexplicable phenomena, and no astronomer looked forward to its return in 1852 with more anxious interest than he. Would two independent comets be found traversing the same path ? or would the phenomenon of a double nucleus be again exhibited ? or would the two components manifest mutual relations analogous to those of satellite and primary, or at least to those of binary stars ? Would it be possible for observations of each component at the coming perihelion passage to be combined with those made at the last return, so that an ellipse could be deduced for each, and the point of intersection thus determined ? These and many similar queries were often discussed ; and immediately on the completion of his paper on the comet of 1843, he began his preparations for an equally thorough investigation of Biela's comet so soon as its approaching return to the sun should have been thoroughly observed.

For a month previous to the detection of the comet, HUBBARD had been engaged in the preparation of an ephemeris to insure its discovery at as early a date as possible, and had succeeded in obtaining an orbit decidedly better than SANTINI's, which was the best existing. But the discovery of the comet rendered the publication of this ephemeris unnecessary.

On the 26th August, 1852, Father SECCHI, at Rome, while searching for Biela's comet in the neighborhood of the place indicated by SANTINI's ephemeris, discovered a very faint nebulous comet somewhat more than $4\frac{1}{2}°$ from the place predicted for Biela's, and was able to fix its position with great accuracy by its transit over a small star of the 9.10 magnitude, which it covered at one time so centrally that the comet could only be recognized by the circumstance that the star seemed enveloped in a faint nebulosity. "I do not know," he adds, "whether this is a new comet or a portion of Biela's which was divided in the beginning of 1846."

There seemed but little room for reasonable doubt that this was really Biela's comet, or one of its component parts; since its position, though varying from the ephemeris, was nearly in the same orbit, and the amount and direction of its motion were what might have been expected. But all doubt was removed three weeks later, when Professor SECCHI detected the other portion of the comet, following its predecessor by about half a degree of right-ascension, and about half a degree farther south, and fainter even than the other. Owing to this extreme faintness of both portions, observations could only be continued for a little more than ten days after the discovery of the second component. The last return to perihelion took place in 1859, but the position of the comet was so unfavorable, that although ephemerides prepared by three independent computers, one of them HUBBARD himself, agreed very closely, and the most powerful telescopes of the world were occupied in the search, the comet was not seen.

With this brief sketch of the history of our knowledge of
Biela's comet, I may, without entering into close detail, describe HUBBARD's labors and researches concerning it. His
published Memoirs on this subject are three in number, in
addition to sundry smaller communications on special points;
such as one in which he corrected a serious error, which had
found its way into the best European computations of the
perturbations in 1845 – 6, and explained its probable origin;
and a publication of the valuable manuscript observations
made by Professor CHALLIS in Cambridge, England, during
the same period, and sent by this distinguished astronomer
to Professor HUBBARD for employment in his investigations.

The first of these Memoirs is entitled, "On the Orbit of
Biela's Comet in 1845 – 6." In this, as in every other memoir of its author, the same searching thoroughness and
scrupulous accuracy are manifest which I have recounted
concerning his investigation on the comet of 1843. All
known observations were employed, no appreciable refine-
ment of method or computation was neglected; and the materials were so fully and completely discussed that it is improbable that any results can ever be drawn from them
which he did not himself deduce. The principal results of
this memoir, in addition to the discussion of all the observations, consisted in the definite determination of elements for
each component, together with their variations for any variation of the adopted mean motion; and in the discovery that
by far the greater part of the difference between the two
orbits might be represented by a variation in the mean
anomaly alone. The residual errors implied by this assumption are very small, much less than the errors of individual
observations, and in no case exceeding 8″; but they are
nevertheless too symmetric, and too large for his normal

places, and he points out, moreover, that some difference must necessarily exist in the mean motions.

In a letter of about this date (1853, June 8), he writes, jestingly: " Biela slides on smoothly. I don't work now, as on '43, wearily and with a $D\psi$, nor boldly and with $D\phi$ance $D\Omega$ing a change of Inclination, but $D\mu$rely. An allowable change of 0."34 in the mean motion will give the places in 1852, within 24" + the error of SANTINI's perturbations, provided I am right in assigning the nuclei relatively to each other; but it is not so easy to tell which is which, as I had supposed."

HUBBARD's published investigations reached this point in the summer of 1853 ; and he was leisurely preparing the materials for a continuance of the work, when the Imperial Academy of Sciences of St.-Petersburg, in December of that year, offered its astronomical prize for just such an investigation as that on which he was engaged. The distinguished head of the Observatory at Pulkowa wrote specially to suggest the publication in the United States of the Programme for the prize ; and it may well be suspected that the very able discussion which HUBBARD had already given might, at least in some degree, have tended to assure the astronomers of the Imperial Academy that competent men were already enlisted in the investigation, whom the liberal prize might at once stimulate and reward. And in view of the laborious and extended computations, which the solution of the problem would entail, a period of nearly four years was allowed for the preparation of the memoir. Many of HUBBARD's friends desired him to compete for this prize, which I think there is no reasonable doubt would have been won by the memoirs which he subsequently published in America.

But HUBBARD's delicate health, together with his earnest

9 *

desire that whatever he might do for science should inure to
his own country's service, prevented him from yielding to
the temptation. He considered the matter for a brief period,
and then decided that he "ought not to work against time,"
and the close of his researches was not reached till 1860.

The second paper, published in July, 1854, is entitled,
" Results of additional Investigations respecting the two
Nuclei of Biela's Comet." In this short, but very elaborate
and important memoir, HUBBARD discussed the observations of each nucleus in 1852, determining elements for
each. And he arrived at the very remarkable results which
seem now incontestable, " that notwithstanding the increased mutual distance of the two nuclei, their alternation,
of relative brilliancy were much greater than those noticed
in 1846; so great indeed, for several days, as to amount
to alternations of visibility from day to day "; and that the
observations at Berlin, 1845, November 29 and December 2,
were of the primary nucleus, the second being invisible to
the observers; while those of CHALLIS, December 1 and 3,
were of the secondary, the first being unseen. So that it is
clear, both that we are in possession of observations of the
second nucleus, made in the beginning of December, 1845,
before the existence of two nuclei was suspected, and that
even at that time occurred those singular alternations of
light which were repeated in 1852. Furthermore, he made
it highly probable that the preceding component, in 1846,
was identical with the following one in 1852, and *vice versâ;*
and finally, that the separation of the nuclei must have
occurred not far from 316° of heliocentric longitude, corresponding to a time about five hundred days before the perihelion passage of 1846.

At the close of 1858, HUBBARD published a short papers
containing a condensed notice of the condition of the prob-

lem, together with new elements for each nucleus, and an ephemeris for each at the approaching return of the comet to perihelion. This I have not counted as one of the Memoirs. His third and last paper on the subject appeared in May, 1860, under the title, "On Biela's Comet." It consists first of an admirable history of all our knowledge of this comet, with full references to the original sources, and presents an excellent specimen of what might be called condensed detail. Next it contains an elaborate discussion of the observations and orbit for every recorded appearance. And in the discussion of the last appearance in 1852, he brought to light a new illustration of the mysterious alternation of brilliancy between the two nuclei. For he showed, that when, on the 15th of September, SECCHI found both nuclei, and determined the position of one of them, the new one being too faint for observation, this so-called "new one" was the identical nucleus which he had discovered in August, and had been observing ever since; while the brighter of the two had then just become visible. "On the 16th, the southern nucleus alone was visible; on the 17th and 18th, only the northern; and finally, on the 19th, both were observed by SECCHI. The double observation was repeated at Rome and at Pulkowa, on the 20th, 23d, and 25th; while on the 21st only the southern, and on the 22d only the northern, was visible. We thus have a most interesting repetition of the alternations in 1845 – 6, which now appear more remarkable only in consequence of the extreme faintness of the comets, which was such, that the slightest change of light sufficed to carry them within or beyond the scope of vision." (*Astron. Journal*, VI. 140.)

Finally, a recapitulation of the final elements for each nucleus, and for all the observations and normal places, exhausts the sum of our present knowledge of Biela's comet, and leaves us ready for the new investigations which its return eighteen months hence will require.

Another extended investigation by HUBBARD is that upon the Fourth Comet of 1825. HANSEN had long ago found that the observations before and after perihelion seemed better reconciled· by an ellipse than by a parabola; and HUBBARD undertook the collection and discussion of all the observations in the hope of some definite determination of the major axis. This investigation occupied much of his time at irregular intervals for five or six years, and was finally published in the spring of 1859. In this, as in most of these cometary investigations, a leading object was to learn whether the motions of the comets, distinguished by their magnitude or varying aspect, or by any other striking peculiarity, would prove in all cases amenable to the law of gravitation alone. In the case of the comet of 1825, no special fact of general interest was elicited; but negative results, though less interesting, are attained with no less labor and skill than positive ones, and are often scarcely less important. Suffice it to say of this memoir, that it is complete, and apparently exhaustive; that the elliptic character of the comet is fully demonstrated, although its periodic time must be exceedingly long; and the material deducible from past observation lies ready for the hands of the future investigator.

I have now spoken, Gentlemen, at sufficient length of the larger and more extended memoirs of our departed colleague, and have described their characteristic features. Of his minor contributions to astronomy I need say no more than that they resembled the larger ones in thoroughness and neatness of conception. The columns of the Astronomical Journal, and the pages of the Washington Observations, are full of them: — elements and ephemerides of many a

comet and many an asteroid, elegant and appropriate suggestions, generally relating to methods of computation, or in-
genious devices for attaining a desired end with economy
of labor.

In the excellent tables appended by Professor COFFIN
and himself to several volumes of the Washington Observations; in the reduction and discussion of the geographical
observations made by Lieutenants (now Major-Generals)
FRÉMONT and EMORY on their various expeditions; in the
thorough investigations of the several instruments successively placed in his charge, — the accuracy and conscientious-
ness of HUBBARD still bear fruit for us.

One of his latest labors was an unpublished investigation
of the magnetism of iron vessels, and its effect upon the compass, — upon which he was employed nearly to the time when
a Permanent Commission appointed by the Navy Department
undertook the same research upon that more extended scale,
which the same gentlemen have continued till the present
time in the form of a committee of the National Academy.

No description of HUBBARD's intellectual character could
be regarded as complete, that omitted one predominant trait
which pervaded all his opinions, and lay deeply rooted in
the very foundations of his nature. I mean that deep love
of truth and loathing of all false assumption, which may be
said to bear the same relation to honesty that honesty bears
to what is called " worldly policy." There were few things
which his modest and tolerant spirit could be said to hate;
but he did hate sham, humbug, and charlatanism with all
the energy of his soul. He never claimed honor, rank, or
position for himself, although he hastened to accord all these
to others far less worthy than he; but he was restive at the
sight of scientific rewards unworthily bestowed by incom-
petent tribunals; and his sterling patriotism and sense of
justice not unfrequently united in paining him, when —

" He saw the holy wreaths of Fame
Profaned to deck ignoble brows."

Thus far, gentlemen, I have endeavored to describe Professor HUBBARD to you as a man of science, — showing you the early efforts of his mind, and the eager pursuit of knowledge which characterized even his boyhood. We have seen what he had accomplished at the age of thirty-nine; and alas! how much more he promised for the future which we hoped for him. But though all this is done, I feel that the more difficult part of my duty to his memory remains undone; and I approach it with yet greater distrust of my ability to do it aright. It is comparatively an easy task to trace the working of his mind, and the results of his studies; but to show him as some of us knew him, as a son, a brother, a friend, a Christian, to do him justice without trespassing on that privacy which none valued more highly than he, requires a hand of equal delicacy and skill. One assistance at least the biographer of HUBBARD may justly feel to be accorded him, — that in that life there is no record to be concealed, no page to be glanced at and quickly closed with pain. His only choice is what to show, not what to hide.

Our colleague had a kindly, gentle nature, and an affectionate regard for all around him. He made his own opportunities to help and cheer others, instead of waiting for them. Was a friend successful, he rejoiced with a cordiality that made him twice happy; in sorrow, he mourned with him, and with a sympathy that half lifted off the burden. One of the strongest affections of his life was for his mother. He showed her not only the natural affection and tenderness of a son, the respect due from youth to honorable age, or the attachment which old and cherished associations awaken, but to the very last he made her his confidante and counselor. His deepest thoughts and highest aspirations, his struggles and his joys, were alike intrusted to her; a precious deposit, which her heart knew how to keep and ponder.

Professor HUBBARD was married at the age of twenty-five to Miss SARAH E. L. HANDY of Washington, on the 27th of April, 1848. Few men were more fitted than he to enjoy the comforts of a home, or could better appreciate the blessings of his new relation; but there were many clouds to overshadow the horizon, as he himself says in one short note, whose pathos only those can understand who know that it was but once or twice in a lifetime that a murmur escaped his lips. Upon the threshold of his home stood always that dreary visitant, Ill-health, whose dominion over both mental and physical content most of us know too well. HUBBARD's own health was never certain, but his wife was a far greater sufferer; and often, unknown to herself, her troubles weighed too heavily upon his over-tasked mind and sensitive heart. Even pecuniary embarrassments, those petty cares that, unlike deeper sorrows, fail to brace the mind they attack, were not wanting to sting his delicate and generous spirit. Each day their peculiar circumstances compelled new outlays, to be defrayed only from means already too slender. We can appreciate their struggles, without prying too closely into what he might have wished forgotten. We can see the student compelled to forego his cherished pursuits, the man of tender sensitiveness wrung by the sufferings of those nearest him, the invalid whose frail health varied with each new trial. We can see all this; but to a spirit such as his must have come many a compensation, many a blessing won from the dark angels by bitter wrestling.

> " For that high suffering which we dread
> A higher joy discloses;
> Men see the thorns on Jesus' head,
> But angels see the roses."

After eight and a half years of married life, a long desired change came to the little household, and with a new
joy he welcomed his child. With what hope and happiness
he accepted the new promise, those who knew him well
cannot forget; but the happiness was all too short. "The
little spirit only fluttered for a while on the threshold of its
prison-house, and unconscious of captivity took flight forever." Writing to a near friend at the time, HUBBARD
says, "God bless you for the interest you took in my boy.
This is all I can say; for I cannot write of him." Nor
will I undertake to speak of his grief. Four years later,
Mrs. HUBBARD's suffering life terminated; and her husband
was left alone, with only the remembrance of a home.

As a friend I knew HUBBARD well, and can bear witness
to the loyalty and gentleness of his nature. With a gayety
never bordering on excess, a sympathy never exhausted, a
kindly tact never forgotten, he was a companion such as we
rarely meet. Of his help and encouragement to me personally, I have already spoken; and since I have read the
memorials entrusted to my care, I see that what he did for
me he did for many others, each according to his need.

During the last few years of our colleague's life, there
seems to have been some modification, or at least exaltation,
of the views and sentiments which, perhaps more than any
others, tend to make each one of us what we are, — I mean
our sense of personal relation to the Deity. That high
principle and religious fervor which through his life had
been a lamp to his feet, showing itself in love to God and
man, burnt during these later years with a yet brighter
flame. Perhaps, indeed, it may to some of us seem for a
moment to have dazzled his vision, and made the shadows
which must darken every thoughtful mind seem blacker
than those ordained by the hand of a loving Father. In

reading these last memorials, we cannot but grieve that his pure and gentle spirit should have passed through those hours of struggle which, to our vision at least, he seemed to need so little. But it is not for us to scan too closely the sacred privacy of these emotions. Let us turn rather to their results.

He was long connected with the religious society of Rev. Dr. GURLEY, in Washington; and his letters to his mother show the reliance which he placed upon this excellent man, and the eagerness with which he sought to know and do his Master's will. He became an elder of the church, and not many months before his death, Superintendent of the Sunday Schools of the Presbyterian denomination in the city. In the affections and lives of his associates and pupils, we find the best tribute to the ability and fidelity with which he discharged these duties.

Among the writings of these later months are various treatises on religious and theological subjects, and critical comparisons and reconciliations of various portions of the Old and New Testaments; to all of which he brought the same power of unwearying research that characterized his scientific labors. He attempted the mastery of the Hebrew language, and labored zealously to fit himself for a more critical study of the Bible. Indeed so earnest was his religious devotion that we find indications of some vague aspiration, or half-formed plan, of renouncing even his scientific pursuits in order to enter upon the labors of the Christian ministry. To each one is intrusted his peculiar gift; and we who knew HUBBARD as a student and minister of science, cannot but feel that his Maker had clearly pointed out the way in which he best might serve Him, by devoting a rare capacity and pure heart to the study and interpretation of His works.

10

Perhaps, Gentlemen, we may regret that, even for a moment, and from the highest possible motives, he was unfaithful to his earliest choice, and swerved from the path where, as we think, he best served God and man. Yet such questions must be decided by every man for himself, with such light as he may attain; and it may be that these varied experiences and changes of thought were sent him that he might live through the experience of many years during the lapse of a few, and might learn as many as possible of the lessons of this life during the time allotted him. But can we do otherwise than honor a creed which blossoms in such deeds as crowded the last years of HUBBARD's life. His was no bigot's zeal. It led him among the poor, the sick, and the afflicted. It sent him to the hospitals, where he daily spent his hours of official leisure with the soldiers, giving each the needful word of good cheer, or bringing delicacies and comforts to them so far as his own opportunities or those of his friends permitted. It inspired him with a true loyalty to his country, and endowed him with that spirit of self-sacrifice which shone in every action. "The number of letters that he wrote for wounded soldiers," says a friend in writing of him, "was almost incredible. He frequently devoted whole afternoons to this one object. I wonder how many of the soldiers knew whose bright face it was that was so pleasant to them."

With all these self-imposed duties added to his daily and nightly routine of work, who can wonder that his health, always so uncertain, became each month still more impaired; and that when the last summons came it was so quickly answered.

Professor HUBBARD left Washington for the last time on the 30th July, 1863. For a few days previous he had been particularly unwell, owing to severe exposure in a sudden shower. But he had looked forward with peculiar pleasure to a meeting of his college classmates in celebration of the twentieth anniversary of their graduation ; and he managed to pursue his original plan, and reached New Haven in time for the meeting. But the delicate instrument had been too much shattered to recover its tone, and its music was to be heard no longer. He suffered severely on the journey, and on being assisted into the well-remembered house where his mother was awaiting him, had only strength to say, " O, how good it is to be at home." His mother pressed forward to meet him, and he added those words, which to our ears seem so full of pathos : " Mother, I am worn out."

And so indeed it proved. To the physician who was instantly summoned, he only said : " Doctor, help me to a little strength to meet my class to-morrow night, and then I will give up." But even this gratification was denied him, and the affectionate greeting that he sent his classmates was almost his last earthly utterance. Gradually, but surely, he sank away ; but who could have wished for him a happier dismissal ? Soothed by familiar voices and pleasant images, tended as in his infancy by his mother, surrounded by loving faces, the worn-out man may have felt himself a weary child again ; and with a childlike confidence he went to rest, on Sunday morning, August 16, — waking, we may be sure, to exclaim once more, " How good it is to be at home." A day or two afterward his mortal part was laid in the quiet cemetery near us, where, two years before, that very week, he had seen his father laid.

> " Sleep sweetly, tender heart, in peace;
>   Sleep, holy spirit, blessed soul,
> While the stars burn, the moons increase,
>   And the great ages onward roll.
>
> " Sleep till the end, true soul and sweet,
>   Nothing comes to thee new or strange;

> Sleep, full of rest from head to feet;
> Lie still, dry dust, secure of change."

Here we leave him. But, gentlemen of the National Academy, let not the name of the first who left our ranks be soon forgotten. Others of those ranks may have emblazoned their names more conspicuously, their memory may be yet more secure of perpetuity, in the annals of science. But none of our number can claim to have surpassed him in those qualities which make the highest glory of a man; and well will it be for us if our names can be inscribed near his, on the highest of records.

If our National Academy is to fulfil its loftiest mission, and achieve a work commensurate with our hope and faith, let us emulate the spirit of him whom we have first been called upon to mourn, — the spirit of disinterestedness, of patriotism, and of highest purpose.

Cambridge: Printed by Welch, Bigelow, & Co.

www.ingramcontent.com/pod-product-compliance
Lightning Source LLC
Chambersburg PA
CBHW021942190326
41519CB00009B/1110